IS FUZZY LOGIC FOR REAL?

A BRIEF INTRODUCTION

BY

LUISA N. MCALLISTER, EMERITUS,
DEPARTMENT OF
MATHEMATICS AND COMPUTER SCIENCE,
MORAVIAN COLLEGE,
BETHLEHEM, PA 18018, USA

Trafford
PUBLISHING

Order this book online at www.trafford.com
or email orders@trafford.com

Most Trafford titles are also available at major online book retailers.

Note for Librarians: a cataloguing record for this book that includes Dewey Decimal
Classification and US Library of Congress numbers is available from the Library and Archives
of Canada. The complete cataloguing record can be obtained from their online database at:
www.collectionscanada.ca/amicus/index-e.html

Printed in the United States of America.

ISBN: 978-1-5539-5882-6 (sc)

Trafford rev. 01/28/2012

www.trafford.com

North America & international
toll-free: 1 888 232 4444 (USA & Canada)
phone: 250 383 6864 ♦ fax: 812 355 4082

CONTENTS

FORWARD

This text is divided into two parts and an Appendix. The goal of part I. is to establish a foundation of fuzzy logic. The goal of the second part is to answer the question in the title in the affirmative because it shows how fundamental fuzzy logic is to the modern treatment of control. Part II also gives details on the famous example of the stabilization of the inverted pendulum by contrasting the olden days solution method with the modern way. Since the aim of this book is to present a concise perspective of the topic of fuzzy logic besides a basic discussion of its fundamental concepts, the references were subdivided into four sets so that one set consists of those references that are cited within, the others provide a guide to a set of recommended reading on several areas, more specifically, they are: (A) On [A] actuarial mathematics; (P) On [PP] popular press references; (N) On [NN] neural networks and fuzzy logic. Finally, there is a set [AP] On applications of a wide variety from computer vision to risk analysis, and, finally [U] On uncertainty modeling. Occasionally, when an application is given at an appropriate time, its references are included immediately at its conclusion. The Appendix is included to provide a brief review of fuzzy sets that constitutes the actual basis to fuzzy logic.

ACKNOWLEDGMENTS

With admiration, love, and gratitude, this book is dedicated to Professor Lotfi A. Zadeh for his teaching, support, encouragement, particularly during these last few years of my recuperation, to my husband and to all those who supported me with friendship and love, to Matt. Kressly for his help with the computer-drawn-figures, and to my dear friends especially Sue Wolfe, Robin Schultz, the Ameglio family, Leslie Hartman, Diane Wood, Mickey Ortiz, and, of course, to all my children.

Luisa Nicosia McAllister
Bethlehem, PA
April 2004

PART I

FOUNDATIONS

CHAPTER 1: WHAT IS FUZZY LOGIC?

Is it the paradox of the century?

How can fuzzy thinking be logical, [4, 6, 8,16]? However difficult this may be to accept, Kosko [5] makes it clear that it has come to be a reality and a new successful way to deal with uncertainty. In particular, we wish to have a computer be capable of reasoning with vague statements that have no probabilistic meaning [13]. This is what we are preparing to illustrate. There is no complete agreement within the fuzzy community on a definitive definition for fuzzy logic. However it seems only reasonable to follow a definition according to L. Zadeh [82, page 78] when he said in a talk he gave at the National Science Foundation that:

♦ "In a Narrow sense, fuzzy logic is a logical system that aims at a formalization of approximate reasoning. As such, it is firmly established in multiple valued logic, but its agenda is quite different from that of the traditional Lukasiewicz's logic because fuzzy logic as logic of approximate reasoning is not part of the traditional multi-valued logic. Although originally there was much antagonism, the scientific community has remarkably changed its attitude. For example, Zadeh has kept a close count, and was happy to announce at a BISC seminar in September1999, that there were 3240 papers, between 1995 and 1998, containing *fuzzy* in its title cited in Mathematical Reviews in contrast of only 521 in 1993.

♦ In a Broad sense, Zadeh often adds, fuzzy logic is almost synonym with fuzzy set theory which, as the name suggests, is basically the theory of "classes without boundaries".

Thus, it is highly capable to handle ambiguities and vagueness.

Finally, most of us ingeniously propose that fuzzy logic is an extension of multiple-valued logic to the continuum case. With the sharp distinction that in the multiple-valued case we are limited to the use of rational numbers within the unit interval, while in the continuum case we may use any real number within the unit interval".

Thornber [16] makes a clear case for this claim. In [12], with the introduction of the linguistic variable, Zadeh makes it even more clear how the tool of fuzzy numbers allows the representation and manipulation of the meaning of vague concepts whose value is not probabilistic, and certainly not stochastic. Why insist that only probability can handle uncertainty when we have long recognized different forms of uncertainty. For further explanation, read Novak, see references [11,12]. Different kinds of uncertainty should be treated and studied accordingly.

EXAMPLE 1. Suppose we say: "*Wanted: a cheap house in a good neighborhood with an excellent school system*".

The vagueness, that is absolutely linguistic, of the above statement, an example due to my good friend Mike Smith, is without a doubt, unsuitable to any attempt at a probabilistic approach, and keep its original meaning.

EXAMPLE 2. Suppose we say *"Matthew is young"*.

A probabilistic expression of the above would change the statement into something like:

"The probability that Matthew is young is 0.9".

Note the implicit law of the excluded middle: Matthew is either young or he is not. What happens is that we only have a 90 % likelihood of being right in knowing what he is Thus, we could have said, using probabilistic language:

"There is only a 90% chance that Matthew is young".

In a fuzzy set context, we would say:

"The membership grade of Matthew within the set of young people is 0.9"

or that, by using any of the permissible fuzzy quantifiers:

"Matthew is **more or less**, [*or* **somewhat**] *young"*.

The semantic is clearly distinct. Again a strikingly distinct semantic that includes quantifiers of a new kind. In other words, the distinction is in the use, in its meaning, thus in the semantic. We shall often have to return to this distinction that is fundamental to remove the antagonism and the confusion often existing between the idea of probability values and degrees of membership because most often they end up to be from the unit interval.

In a fuzzy context, we have a continuum of possible valuations which means that mathematically we can use anything within the unit interval. If we were able to say that the value of the truth is either zero or one, then we do not hesitate, and we have a crisp case. It is when we are unsure that the truth-value is neither 0 or 1, maybe it is somewhere in between, then we may believe that it is a likely event, then we have a probability, when it is not this case either then we have a fuzzy case and thus we cannot force a crispness or probability modeling approach. Thus we must interpolate. Note the frequent mention of the term **Interpolative** in the quote below. It turns out that these matters lend themselves to the use of fuzzy numbers. Finally, we conclude this introductory section by quoting a concise viewpoint as expressed by Zadeh at his NSF talk in May 1993. Namely, he said:

"The basic concepts of fuzzy logic are:

♦   The concept of linguistic variable, that is, a variable whose values are words rather than numbers;

♦   The concept of canonical form which represents information as an elastic constraint on a variable;

♦   The concept of **Interpolative** reasoning which makes it possible to reason with incomplete information.

♦   **Interpolative** reasoning plays a key role in human cognition and lies at the basis of pattern classification, qualitative reasoning, system identification, system modeling and neural network modeling. In the context of fuzzy logic, a concept which is central to **Interpolative** reasoning is that a collection of fuzzy if-then rules which serve to provide an approximate characterization of the input–output relation of partially specified system approximation. To fill the gaps in knowledge, a number of different architectures may be employed, prominent among which are disjunctively combined fuzzy if-then rules with a defuzzifier and the Takagi-Sugeno-Kang fuzzy if-then rules with a convex aggregator. The role-played by **Interpolative** reasoning in the implication of fuzzy logic. The importance of fuzzy logic derives from the fact that almost all of human reasoning is approximate in nature. In fact, it is the ability to recognize distorted speech, decipher sloppy handwriting, and, more generally, make rational decisions in an environment of uncertainty and imprecision". Note that the membership function is not subject to any restrictions other than to be continuous on the interval of interest.

♦   **EXAMPLE.** To emphasize the distinction between probability and the concept of membership function, consider the case in which a patient reports abdominal pains to a doctor, and she is then asked to quantify the severity of the pain on a scale from 1 [no pain] to 10 [very high]. There is no probability involved, the pain exists, its severity is the only number that has any meaning.

♦   Personally, I believe that a good intuitive definition is to say that fuzzy logic is an approximation to classical logic just as numerical analysis provides an approximation to exact solutions.

CHAPTER 3: ABILITY OF FUZZY LOGIC OF HANDLING FUZZY QUANTIFIERS

This major ability is what sets it drastically apart from any logic that has been studied previously. In [17, page 13 and following], Yamakawa, a famous Japanese scientist, gives a clear instance of the use of fuzzy numbers for the modeling of complex systems. He makes a parallel between a mathematical formula and a linguistic rule. This rule would be expressed in a way that is something like the one below:

RULE i: "If x is A_i, then f (x) is B_i".

where A_i and B_i are to be interpreted as fuzzy sets or linguistic constants. In fact, a relation between a variable x and f (x) can be stated by a set of linguistic rules or by a mathematical equation. Linguistic constants are generally well-defined languages, namely either crisp information or exact numerical values in terms of crisp rules, we have slightly different if-then clauses:

EXAMPLE 3. For $f (x) = [(x - 3)^2] \div 2$; then

RULE 1: if x is around −2 then f (x) is around 25/2.

In the same paper, continuing his summary, he adds that a set of rules forms a knowledge base. The inference system cannot contain contradictory rules. In [17, page 16], he gives examples of linguistic rules with ill-defined language such as rule 1 above or as:

RULE 2. *"If the temperature grows higher then the opening of the valve should be reduced significantly, and the fuel should be reduced a little".*

Note that many cooking, and driving, instructions are quite similar to rule2, yet most people succeed in making sensible interpretations. can we achieve enough ability to enable a computer to make sense out of these instructions? Knowledge bases constructed from experts are rules that are hardly ever precise. The incredible success in using fuzzy logic to solve control problems has many scientists wonder whether teaching differential equations has become obsolete. Balancing the inverted pendulum leads to a complex system of differential equations. Not so, if the control has a fuzzy logic rule base. The ease of such technique is clearly appealing, as it will be demonstrated in section 9. Likewise, if we wish to balance a moving platform with a glass of wine and a mouse on it. To conclude, why take our measurements to be exact when they are so rarely that way? And as a final

EXAMPLE 4:

When we say *"Kathy is six feet tall"*, we do not understand "She is about six feet tall". Rather we understand that she is exactly six feet tall.

Likewise, if we say *"Dr.Dara is possibly mid-sixty"*

We do not take that to mean that she is exactly 65 years old. We correctly discern that the value as an approximation, not as an exact value. Thus, we should treat these values as imprecise rather than exact. A well-known interpretation of fuzziness is that just as numerical analysis deals with the approximation of problems that are so complex that we cannot find the solution in a closed form, so fuzzy sets and fuzzy logic respectively deal with an approximation of crisp sets and of truth-values. Keep in mind the parallel between numerical analysis enabling us to find solutions to difficult problems in an approximate fashion and fuzzy logic, giving us approximation methods to find solution to communication problems. Why always cling to an illusion of exactness when it is not often needed and cannot be used either.

CHAPTER 4: DIFFERENT KINDS OF NUMBERS.

RossT. [AP26, page165]

Interval arithmetic can be thought the following way. When we add, multiply any two crisp real numbers, the result is still a single crisp real number, or a singleton. If we combine an infinite number of crisp singletons from an interval,we expect a crisp singleton. Thus, an interval is the result from combining interval.

The topic of interval arithmetic has found its proper place in modern mathematics. Its application to imprecise sentences is quite natural. Everyone has the correct understanding of a sentence such as:

"*They are a pair of thirtysomething people*".

Nobody expects that to mean other than their age is something between t=30 and t =40, namely the value of t belongs to the interval [30, 40]. In other words, the interval becomes a representation of the imprecision of the sentence. What we do now is to introduce other methods to establish a correspondence between an imprecise sentence and a mathematical expression.

What is our goal? In this discussion we accept realistically that often we need to use numbers that <u>approximate</u> closely our evaluation of an item or even of a situation when we cannot do so <u>exactly</u>. Everyone agrees that to assume or make use of those approximations as if they were exact is conducive to nonsense. This is quite natural, and in physics we are taught early on that every instrument has an error associated with the measurement it yields, thus, we must be aware of it and include it in our calculations. To accommodate this line of thinking there have been attempts to method probabilistic reasoning, see references [U 1,11, 19], none with great success, check and read the various works in the section of the references on modeling under uncertainty, [U], page47. Thus, it is desirable to conserve clearly the character of imprecision of the evaluations for example by creating in a suitable fashion numbers that reflect the imprecision and then manipulate these numbers with much savvy. In the next chapter, for example, **mid-sixty** can be represented by a triangle with vertices (60, 0), (65, 1), (70, 0).

Note that 65 has been associated with a value one because it is truly mid-sixty, while sixty and seventy are associated to the value zero because they are clearly not mid-sixty. The latter representation is referred to as a **Triangular Fuzzy Numbers**. Incidentally, there exists a well formulated fuzzy arithmetic that some believe it will soon included in hand calculators [9]. Other fuzzy numbers, very popular in applications, have the shape of a trapezoid, or exponential.

Trapezoidal Numbers are also very popular in the representation of imprecise expressions. Buckley uses them successfully to generate a computational scheme to analyze financial matters [23]. Their construction is similar to that of triangular numbers as it will be illustrated in the later example, next chapter, example 6, page 24. Arithmetic that handles either numbers is available and it will be included in later sections, see Buckley's paper [23], where he proposes a nifty technique, or a recent text by Kauffman and Gupta [22], and others, e.g. [1]. As a matter of fact, some are already predicting the inclusion of fuzzy numbers in hand held calculators [9].

NOTATION. The trapezoid with vertices (a,0), (b,1), (c,1), (d,0) is denoted concisely by T **[a, b, c, d]**.

Where a, b, c, d are any real numbers of the horizontal axis, and these vertices are joined by segments of a line in the order they are given. See example 6 in the next section.

CHAPTER 5: WHAT IS A LINGUISTIC VARIABLE?

We quote an accepted definition [19, 20].

DEFINTION 1. We say that X is a **Linguistic Variable** if its values which are called interpretations, are natural languages expressions [AP19]. As an additional comparison, in classical logic there are two quantifiers, the existential and the universal. Not so in fuzzy logic, some of which are:

$$\{almost, rather, often, approximately, close\ to,\}\ \ldots\ldots\ (3)$$
$$\{near, few, later, several\}\ \ldots\ldots\ (3a)$$

It is no wonder that mathematics must become fairly sophisticated. Zadeh [18] applies the concept to approximate reasoning that is at the core of the effort of making machines achieve some form of reasoning, much like humans do. In fact, the ability of a human to understand slur speech and fill in details for some missing information is amazing. The valuation of a classical logic implication is rather straightforward, not so for the fuzzy logic case. Many solutions have been proposed over the past two decades with none becoming standardized. With the final selection of a method being primarily due to personal preference. Some texts do a splendid job in examining many options, see Ross [13], Klir's [3,4], and Asai&Terano &Sugeno [1].

One of the most interesting issues still strongly debated within the fuzzy community is how to compute the validity of an implication. A comparative review here is beyond the purpose of this concise presentation, thus we refer the reader to consult the authors that were mentioned above. Finally, it is time to consider, see [2,18].

For the example in section 9, the stabilization of the inverted pendulum, the needed linguistic variables that describe the state of each variable are fuzzy numbers in a triangular shape, see here figure5 are:

$$\{AZ= Approximately\ Zero, NS=Negative\ Small, NL= Negative\ Large,\}\ \ldots\ldots\ (4)$$
$$\{PS=Positive\ Small, PM= Positive\ Medium, PL=Positive\ Large\}\ \ldots\ldots\ (4a)$$

And their interaction appears in the rules matrix in figure4, page46, given by their first letter, where each is a fuzzy set in the shape of a triangle shown in figure5.

CHAPTER 6: WHAT IS A FUZZY SET?

Ross quote [AP26, p.13]

There exists the clearest distinction between fuzziness and randomness.
While the former describes the ambiguity of an event, the latter describes the
uncertainty in the recurrence of an event.

Although we wish to simply mention some basic facts here, we invite the reader to use some of the better texts that are listed in the references such as the Klir and Bo Yuan, Asai etal, Ross, McNeil and Thro, we need to remark that this concept is a slowly growing in understanding and in popularity because it really helps to formulate clearly our idea of a set without sharp boundaries. For example, just a decade ago, we would find perhaps a hundred or so papers in print on the topic, now there are more than thousands. More details on fuzzy sets will be given in the Appendix on page 40. It will suffice here a brief perspective that is needed only because many of us in the fuzzy community, consider fuzzy logic as the same as the theory of fuzzy sets from which it is derived. Note that often we shall juxtapose the terms sharp and crisp to that of fuzzy. Thus, a by ordinary, classical set it is mean a set that is sharply defined because we can easily determine whether or not the its characteristic property is or is not satisfied without hesitation. In contrast, we have a fuzzy set whenever the concept of belonging is not chrystal clear, perhaps because the defining property is only partially satisfied. One of the most powerful tools is the Extension Principle that we use to fuzzify classical mathematical concepts, as Yager aptly explains in reference [26].We shall begin from the basic definition, i.e.

DEFINITION.1. Given a **support** set denoted here as A, normally an ordinary crisp set, the set of pairs

{ (t, $0 \leq m$ (t) ≤ 1} with t in the support A, is called a **Fuzzy Set** denoted A (f) with support A, and membership m (t) normally is a continuous function on some interval [a, b]. Its notation is generally where f reminds that we have generated a fuzzy set from a sharp set A:

$$A (f) = \{a/a \in A, m (a) \in [0,1] \}$$

For even more details, see, or any of the texts that are cited in the references [1,3,9,10,12, 13,13a]. Two types of membership functions have been discussed in the literature.

(I) The <u>discrete</u> case where to each element we associate a real number from the unit interval [this is found in the original Zadeh's definition]; and

(II) The <u>continuous</u> case, [31], whereby m (a) is any continuous function over the interval of interest.This is the most popular, although not the original, definition.

Operations for fuzzy sets are obviously expressed in terms of their membership functions with their result equal to the result we would have should the fuzzy sets reduce to crisp set, whose membership is either zero or one. In its original treatment, Zadeh considered the discrete case whereby a finite number of pairs (the element and its membership) were assigned. His original definition for union and intersection of fuzzy sets were respectively the maximum and minimum operators that are still widely used because of their simplicity. in spite that many suggest their replacement with T, and S- norms.

A very important concept is that of fuzzy number. To consider it, we must introduce some other basic ideas such as

DEFINITION 2. A fuzzy set is called **Convex** if for all s in the interval [u, v] we have that the membership function satisfies the condition:

$$m(s) \leq \min[m(u), m(v)].$$

EXAMPLE 5. When we say the a friend is mid-sixty, we do not necessarily mean sixty-five, we mean somewhere between 60 or 70. Thus we can construct a triangle with vertices (60,0), (65,1), (70,0) where we note that a membership equal to zero is assigned to both 60 and 70 because neither is clearly mid-sixty, however note that the membership equals one for 65 which is certainly mid–sixty. To further comment on how fuzzy numbers are feasible to portray imprecision, consider again the case of the sentence

"*Johanna is about six feet tall* ".

We do not take it to mean that her height is precisely six feet, but that the number that is given is an approximation of her height. To suggest how we feasibly portray a statement by the geometric shape of a trapezoid, we include the following.

EXAMPLE 6. Suppose we wish to asses the performance of a friend knowing that it varies during the day. We may use a qualitative assessment, far more consistent with our valuation. For instance, we know that at beginning of the skiing day, our friend is rather poor, and that after a 20 minutes or so, our friend improves enormously, reaching an excellence in her performance. Using a (t, y) coordinate plane we find that relevant times are t=0,{when improvement begins}; t= 20,{when excellence, i.e.y=1, begins}; t=40, {when ability begins to decrease}; and t=100{when our novice skier is a disaster again, i.e. y=0, }.Do we have the vertices (0,0), (20,1), (40,1), (100,0)? If we label them respectively by A,B, C, D then we draw a trapezoid with sloping sides AB,CD respectively showing increase, and decrease in ability as we had noted. Thus, once again placing a one-to-one correspondence between a geometric shape and the assessment of an athletic performance that varied over time intervals. We might even add that it is far more realistic than any verbal or numeric assessment we may think of at the end of the skiing day for our novice skier. This is illustrated in the next

FIGURE 1. If we wish to compute an evaluation of this skier performance at some time between the time t=0 and t=20, we find the equation of the line joining the point A to B.Thus,

we plot the points and then we connect them. For example, we find the y value corresponding for example at t=10, or y=0 the same holds at t=100. But at t=20 and at t=40, the value ought to be y=1 because then the performance is excellent. Thus, the significant points are:

$$A\ (0,0),\ B\ (20,1),\ C\ (40,1),\ D\ (100,0).$$

For any other intermediate time. Namely, if we want an approximation of an evaluation at t=70, we find the equation of the line CD, and compute its y-value for t=70.

Note that this a nifty consequence that is a **bonus** because given the quality of observation, these intermediate valuations yield rather meaningful approximations of the performance evaluation rather than produce a numerical guess.

SOME CONCLUSIONS

Why believe that we have exact measurements when we do not? At this time, it is feasible to mention another frequent criticism, namely that membership functions are probability functions. Nothing could be more false. What the membership functions can be at most are possibility distributions. Buckley makes use of trapezoids in decision making besides providing an algorithm for a fast use, see [AA5]. In soft computing, a new computational field of research with the goal of computing with words rather than numbers because most researchers conjecture that human beings tend to use verbal assessments generally in a more consistent way than numbers, that are often randomly chosen. Thus, our encounter with triangular and trapezoidal numbers has pointed to a new direction, that that in fuzzy set theory id formally called as we discussed next. However, as it is often necessary to do in mathematics we must define basic terms.

DEFINITION 3. A **fuzzy number** is a **convex** fuzzy set.

Major examples of fuzzy numbers we have already introduced are the triangular and trapezoidal numbers.

It remains to consider the arithmetic of an example.

CHAPTER 7: INTRODUCTORY ARITHMETIC OF A COMMON FUZZY NUMBER

Suppose we receive the following **message**:

"*Mr.T. said about twenty years ago that Dr.Dara was mid-sixty*"

It is feasible to make the following **conclusion**:

"*Dr.Dara is mid-eighty now*".

How can anybody claim that the uncertainty here is probabilistic?.

We have an approximation due to imprecise information If we can reach a conclusion, should an intelligent system reach the same conclusion? Using triangles. The question becomes what is the base of the resulting triangle?

McNeil&Thro provide a technique in their book on a practical approach to fuzzy logic, see [1] in section2. Unfortunately, their explanation leaves much to be desired. The notation for a triangular fuzzy number though is neat.

EXAMPLE 7. Let the triangle have a base beginning at x =-3, and end at x =7, reaching the value one at x = 4, then it is denoted T [-3,4,7]; This notation ought not to cause confusion with that for the trapezoidal numbers we saw previously.

The meaning of T [-3,4,7] is "*about four*"

The value x=4 is called "**Apex**".

As a consequence, all operations, it does not matter whether sum, difference, multiplication, or quotient, are executed on the apex, as in normal integer arithmetic. Why? Suppose you heard someone say a message as we quoted at the beginning of this section does it not make sense to add the 20 years to the possible 60? The real difficulty lies in finding the base for the new triangle with apex equal to 80. Note that according to the notation that was introduced above T [a-c, c, a+c] is a trapezoidal number whose membership is then given by $y = 1- |x -a| / c$. Using $y =0.5$, we solve for x the equation

$|x -7|=2-1 (0.5)$, namely $|x -7|=1.5$, as we have above. Thus, the solution of $x -7 =1.5$ and of $-x+7 =1.5$ are respectively $x = 7.+1.5=8.5$

and $-x=-7+1.5=$ '- 5.5 so that $-x=-5.5$yields $x=-5.5$ we have generated a base equal to the interval [- 5.5, 8.5]. It remains to review the basic operations of interval arithmetic as used in [8], as we list next. Assume $a \leq b$, and $b \leq c$ then we have:

Addition: [a,b] + [d,c] = [a+d, b+c];

Difference: [a,b] − [d, c] = [a-d, b − c];

Multiplication: [a,d] • [d,c] = min (ad, ac, bd, cd), max (ad, ac.bd,bd,bc];

Division: [a,b] / [d. c] = [[a, b] • [(1/d), (1/c)];

See Morore R.E.'s introduction to interval analysis texts.

Where • denotes multiplication and (1/c), (1/d) are the reciprocal of c and of d, therefore both d and c are >0.

Scalar product: u [a,b] = [ua, ub], if u >0;

= [ub, ua], if u<0.

Studies on interval arithmetic abound. They satisfy the curiosity for the use of partially ordered sets see for example Moore's text [28].

How do we extend the concept of mapping to fuzzy sets? The Extension Principle in fuzzy sets is discussed in an excellent way in [27]. Since, for the clarity of exposition, it is not necessary to use currently this principle, we then we move on to other important concepts. Having defined a fuzzy proposition, then we wonder how we combine them in a systematic sensible way. Note that we are trying to establish a calculus of fuzzy propositions, just as it is done in classical logic.

CHAPTER 8: THE COMPOSITION OF FUZZY RULES

In this chapter first we ought to ask why we need fuzzy rules. Then briefly report on results that exist on the compositions.

Note that in classical logic, we have a well-defined prepositional calculus. Thus, we expect to have a similar case developed in fuzzy logic for fuzzy propositions.

First we observe that we have cases of fuzzy propositions that our human brain can handle rather easily. The question we ask is then: can we establish well-formulated rules that make sense for fuzzy propositions?

There i is much work that is available in the literature. We shall report a few next recognizing the name of those who first proposed the formulation that is presented. How about a com- parative study? In which case, see Reference [17].

The following are just a sample

$T(x, y) = \max[1 - xoy]$ ....Kleene-Dienes;

$= 1 - x + xoy$............Reidenbach;

$T(x, y) = \min[1, y - x]$................................Lukasiewicz;

$= \sup\{ t / \min(x,t)\_ y\}$..........Godel's;

$T(x, y) = y$, if $x = 1$, [17][17]

$= 1$, otherwise............Goguen's...[J4];

$T(x,y) = \max[1 - x. \min(x,y)$..........Zadeh's...[J5];

$T(x, y) = \{x \text{ AND } y)OR(1 - x)$.................Mamdani's....[J6] ;

The latter two, [J5, J6] are rather popular in applications to control problems.

However, some claim that [J6] is controversial because it has been shown that it reduces to the Cartesian product, see [6].

Zadeh's original definition [J5] is very popular, often simply referred to as the MAX-MIN Rule, particularly be- cause of its simplicity and giving sensible results [19, 20]. The question that is posed is that we have to find an evaluation for the implication of the type

"If P then Q" ....(J11);

Where we assume that its truth evaluation cannot be clearly established. To explain further what the problem really presents here , suppose that P is the statement "the color is red".

If what we actually mean is that what we have is an approximation to a shade of red, then we have a fuzzy set, denoted A(f), so that if u is an element of the universe of colors, then u has a membership a u in [0,1]. This membership represents a method for valuations, or assessments of the degree of belonging to this special shade; in this example; it assesses what is the level of redness of the item.

Therefore, in what way do we compute the valuation corresponding to the implication in terms of fuzzy sets as expressed by the implication? see [18]. The same reasoning applies to the statement Q., whose imprecision gives way to another fuzzy set, now denoted B(f), with membership m b in[0,1].

A good discussion in reference to this problem appears in [13, p.209] where the author not only cites the name of the person who first suggested the rule but it includes references to the specific work of that author, an important information for anyone who wishes to pursue the matter further. Exciting extra reading is found in the references [1, 2, 3, 6, 8, 13], given next.

The Use of the fuzzy Modus Ponens. Recall that if A, B, C are fuzzy sets that represent linguistic states of a variable. For example for our pendulum problem in the next chapter, we have:

RULE1. If (e, De) is A x B, then v is C;

RULE 2. if (e, De) is A x B, then v is C.

EXAMPLE1. ;

RULE 1. if x is about 4;

RULE 2. IF f(x)=2x then f(4) is about 8.

The use makes much sense; Suppose we say for

EXAMPLE2 :

Rule1. x is about 5;

Rule 2.If f(x) =3x;

then f(x) is about 15.

where A is the fuzzy set of all numbers close to 5, and C is that of all numbers close to10 or 15 for the second example.

Note that one method that has not been mentioned here is the traditional initial phase of fuzzification, see[4, pp.476-479] for the variables, and the final reverse process that is called defuzzification, see [2, pp.336-338]. For an application to cost effectiveness, see [AP24], a work sponsored at the Aberdeen Proving Ground Research Laboratories.

The composition of rules can be obtained by a variety of methods . Most popular in practice are the Zadeh's and the Mandani's rules., see [1, 2, or 4, 8]. For the purposes of most knowledge bases of a fuzzy controller, a matrix representation of each rule is simple and easily read as usual, see the discussion on page 340 of the Klir&Yuan text [3, pages 280&following].

REFERENCES:

First, we list in alphabetical order the references that are actually used in the text; then, we give a list, that begins on page 30, of different interesting references that make good additional reading, including actuarial work,popular press, neural networks and logic, uncertainty modeling, and applications e.g. to computer vision, uncertainty modeling.

[1] Asai K. &Terano T. &Sugeno M. Applied Fuzzy Systems, AP Professional, NewYork, 1994.

[3] Klir G.& Bo Yuan, Fuzzy Sets and Fuzzy Logic: Theory and Applications, PrenticeHall, 1994.

[4] Klir G.&Folger T. Fuzzy Sets, Uncertainfy, and Information, Prentice Hall, N.Y.,1993.

[5] - Goguen J. "The Logic of *Inexact Concepts*", Synthese, vol.19, pp.325-373, 1969.

[6] Kosko B. Fuzzy Thinking:the New Science of Fuzzy Logic, Hyperion, New York, 1993;

[7] -Mamdani E.H., "*Advances in Linguistic Synthesis in Fuzzy C10ontrollers*", Int. Journ.,Man,Machine Studies, vol.8,pages 669-678,1976.

[8] Mc Allister L.M.N.,"*A Computationally Simple Model to Represent and to Manipulate Linguistic Imprecision*" Intern. Journ. of General Systems,16, (3), pp.265-289, 1990.

[9] —McAllister L.M.N."*Can you Ski?*" Math. Magazine, pp.287-295, November 1985.

[10] —McNeil&Thro,Fuzzy Logic,AP,1994;.

[11] -Novak V.Fuzzy Sets and Their Applications,Adam-Hilger, BrisTol,1989.

[12] —Novak V.-"*On the Syntactico- Semantical Completeness of First-Order Fuzzy Logic* "*part I-Syntactical; partII Aspects: Main Results*",KYBERNETYKA, 26, 1990, pp.47-66; pp.134-154.

[13] —Novak V."*Linguistics and Fuzzy Control*", BUSEFAL, pp.87-94, summer1993.

[13a] -Ross T.Fuzzy Logic with Engineering Applications, McGraw-Hill,1995.

[18] Smets P.& Magrez P. "*Implication in Fuzzy Logic*" IJAR, 4,1,pp. 67-78,1986.

[15] —Smucker J.A. *Fuzzy Sets, Natural Language Computations and Risk Analysis*, CRC,1986.

[16] Thornber K.K "*A Key to Fuzzy Logic Inference*", IJAR, 8 (2),pp.105-121,1993.

[17] Yamakawa T. "*Stabilization of the inverted pendulum by High -Speed Fuzzy Logic Controller Hardware System*" IJFSS, 32, 2, pages 161- 180, 1989.and

—— [17a] —"*Fuzzy Logic Controller*", Journal of Biotechnology, vol.24, no.1, pp.1-32, 1992.

[18] Zadeh L.A."*Concept of Linguistic Variable and its application to Approximate Reasoning*" I, II, III, Info Sci., pp.199-249, pp. 301-357, pp.43-80,1975.

[18] -Yager R.R. "*On the Implication Operator in Fuzzy Logic'*, Information Sciences,vol.31, pages 141-164,1983.

[19] Yager R.R "*Implementing Fuzzy Logic Controllers using a Neural Network Framework*"IJFSS, 48,1, pp.53-64,1992.

[20] Zadeh L.A."*Quantitative Fuzzy Semantics*",Info.Sci., 3,1973, pp.159-176.

[22] —Zadeh L.A. *"The Birth and Revolution of Fuzzy Logic "*:, I.J.of General Systems,17 (2-3), pp.95-105,1990.

[22] —Zimmermann H.J.Fuzzy Set Theory and its Applications, Kluwer Academic,Amsterdam,1985.

[23] —KaufmannA.&Gupta M. Introduction to Fuzzy Arithmetic: Theory and Applications, Van Nostrand, New York,1988.

[24] —Buckley J."*The Fuzzy Mathematics of Finance*" IJFSS, 21, 3, pp.139-147, 1987.

[25] -Wang P. Computing with Words, Wiley-Interscience, New York, 2001.

[26] Wang P. (ed.) Advances in Machine Intelligence and Soft Computing,vol. IV, Duke University,1997.

[26] Yager R.R *"A Characterization of the Extension Principle"*IJFSS,18, 3, pp.205-217,1995.

[27] Moore R.E. Methods and Applications of Interval Analysis,SIAM Studies in Applied Mathematics, 2,1979.

[28] Moore R.E. Interval Analysis, Prentice-Hall, Englewood Cliffs, N.J.,1966.

[30] Polya G."*Mathematics and Plausible Reason*ing" Princeton University Press, NewJersey, 1954.

[31] DuboisD.&PradeH.Fuzzy Sets and Systems: Theory and Applications Academic Press, New York,1980.

N.B- For simplicity the following acronyms have been used throughout for:

IJFSS= Journal of fuzzy sets and systems; and for

IJAR = International Journal of Approximate Reasoning.

SPECIALIZED BIBLIOGRAPHY:

ACTUARIAL WORKS:

This is an avant-guard field already offering many possibilities for far reaching results to those who are daring to explore and pursue the lead of these works. The authors of references [from A3 to A5] ought to be contacted for many worthy suggestions.

[A1] Ostaszewski K."*An Investigation into Possible Applications of Fuzzy Set Theory to Actuarial Sciences* "Society of actuaries,Schaumburg, pp.23-67, 1993;

[A3] OstaszewskiK.&Karkowski W. "*Towards Fuzzy Actuarial Credibility*'" NAFIPS PROC., pp.217-221, August 1993 (L. N.Mc Allister,ed);

[A4] Carreno L.A. &Steel R.A. "*A Fuzzy Expert System to Insurance Risk Assessment Using Fuzzy CLIPS*"NAFIPS PROCEED., pp.202-211,1993.

[A5] Karkowski W.&kK. Ostasewski" *New Method is Fuzzy*" The Actuary, May1992.

[PP] POPULAR PRESS:

It is amazing how much understanding has been shown by those who are willing to set aside preconceived notions and listen to basic arguments and explanations. Reading these works is as good a place to start as any.

[PP1] Frederick J."*Comments on Stocks.mart.com*", MoneyNewsline, Agust 1997,pp.33-35,1997.

[PP2] MMcKanK.& Dwoteretsky T."*Fuzzy Means to Logical Ends*" New York Times,11October,D4,1984.

[PP3] Marbach etal: "*Reasoning with Fuzzy Logic*", Newsweek,28,January1985.

[PP4] Arstrong Larry "*Software can Dethrone Computer Tyranny*"Business Week, April 6,pp.90-91, 1993.

[PP5] Kosko B. & Isaka S"*Fuzzy Logic* "Scientific American, pp.76-81, July 1993.

[PP6] —Boyd J."*Fuzzy Frenzy is far from phased out*",The Japan Times, Monday,May24,1993.

[MOF] MATHEMATICS OF FUZZINESSS.

When L.A. Zadeh first proposed the theory of fuzzy sets, his theory was, if nor scoffed, certainly ignored and almost rejected, by many within the American mathematical community. Fortunately, abroad, many immediately caught on with its beauty and challenges, and pursued the topic vigorously, contributing to the development and expansion of an exciting new field of research.

One of the major misunderstandings continued to be the confusion on the notion of probability and of membership grade that is clarified once it is made clear that the difference of the semantic between the probabilistic valuation and the fuzzy valuation is enormous, thus the only common feature is that both probability and membership degrees share, is the unit interval! A rather modest overlap, indeed! The works that are cited here provide a good start for anyone to enjoy a promising, and challenging field with much to do and enjoy. SO… happy reading !

[**MOF1**] Sugeno M."Theory of the Fuzzy Integrals and its Application", Ph.D Dissertation,Tokyo Institute of Technolgy,,Tokyo,1974.

[**MOF2**] BaroneJ.M."*Fuzzy-Relation Preserving Maps and Regular Fuzzy Topological Spaces* ""Proceedings of the NAFIPS Confer., pp.286-294, Syracuse, N.Y., Sep.1997.

[**MOF3**] —Kerre E." *Fuzzy Topologizing with Preassigned Operations*" International Congress for Mathematicians, Helsinki, pp.61-62,1978.

[**MOF4**] Mordeson J. &Wierman M.J. "*Differentiation of fuzzy functions* "Proceedings of the NAFIPS Conference, pp., Syracuse, N.Y.Sep1997.

[**MOF5**] Lowen R."*Convergence in Fuzzy Topological Spaces*", GeneralTopology,and its Applications,10,pp.147-1289,1979.

[**MOF6**] Lowen R"*Fuzzy Neighborhood Spaces*" IJFSS, 7,pp.165-289,1982.

[**MOF7**] Kaufman A. &Gupta M.An Introduction to Fuzzy Arithmetic.

[**MOF8**] Vivona & Benvenuti M.J.&Vivona D."*General Theory of the Fuzzy Integral*",Mathware and Soft Computing

[**MOF9**] Sugeno M. "*Fuzzy Integral and its Application*, Ph.D Dissertation, Tokyo Institute of Technology, Tokyo, Japan, 1974.

[**MOF10**] —Kerre E.&DaBaets B."*On the Supercomposition of Subcontinuous and Fuzzy Mappings and Related Topics*", Proceed. of the NAFIPS Conference,Allentown, PA,pp.232-241,1993.

[**MOF11**] Lowen R. "*Convex Fuzzy Sets*", IJFSS, 3, 3, pp.291-310, 1980.

[**MOF10**] Sugeno M. "*A Survey of the Theory of Fuzzy Control*",Information Sciences,36,pp.295-298,1997.

[**MOF12**] WiermanM.J."*Total Extension of Set Functions and Relations*, Proceed. of theNAFIPS Conferenc.,Syracuse, N.Y.,pp.292-294,1997.

[**MOF13**] McAllister L.N. "*Polynomial Fuzzy Numbers*", BUSEFAL, volume87, 2003.

[**MOF14**] McAllister L.N. "*Fuzzy Graphs and Networks Repairs*" Intern. Journal. of Comput.in Mathematics, vol.80. no.9. pp.1-6, Nov.2003.

[**MOF15**] Rosenfeld A."*Fuzzy Graphs*" in Fuzzy Sets and Their Applications to Cognitiveand Decision Processes, Academic Press (Zadeh L.A.,.&Tanaka K,&Fu A.K.S. &Shimura M., Eds.) 1975.

[**MOF16**] DuboisD&PradeH."*Towards Fuzzy Differential Calculus*", IJFSS, 8, pp.1-17, 1982.

[**MOF17**] Turksen I.B.*Four Methods of Approximate Reasoning with Interval-Valued Fuzzy Sets*" IJAR,3, pp.121-142, 1994.

[**MOF18**] Yao YY."*A Compaison of two Interval-Valued Probabilistic Reasoning Method*", Proceedings of the 6th International Conference on Computing and Information, 1, pp.1090-1105, (paper number D6), 1995.

[NN] NEURAL NETWORKS AND LOGIC:

Much interest has recently been given to neural networks. Some success in their application has been achieved Thus a lot has been written about them. In the references, that have given here there is much interesting to be learnt. Note that some, e.g. [NN4, 5, 6, 7], tend to compare them to fuzzy logic control.

[NN1] Yager R.R. *"Connectives and Fuzzy Quantifiers in Fuzzy Sets"*, IJFSS,vol.40,pp.39-79,1991.

[NN2] Takagi H.&Hayashi Y. *"NN-Driven Fuzzy Reasoning"*, IJAR,5,pp.191-212,1991.

[NN3] Bortolan G."*Fuzzy Networks for Linguistic Processing*"Procedings NAFIPS 1993,pp.42-46,1993.

[NN4] Pedrycz W."*Fuzzy Neural Networks and Neuro Computations*" IJFSS, 56, pp.1-28.

[NN5] IsikC.&Zia F."*Organizing Neural-Net Approach to Fuzzy Logic Control*" Proceedings.NAFIPS, pp.180-184,1993.

[NN6] Chin-Teng&Lee C.S."*Neural Networks and Fuzzy Systems*" IEEE Transactions on Computer Special Issue on Artificial Neural Networks, 40, 12, pp.123-1336,1991.

[NN7] Hayashi Y. &BuckleyJ. &Cgolala E. *"Fuzzy Neural Networks with Fuzzy Signals and Weights"*, International Journal of Intelligent Systems,8, pp.527-538,1993.

[NN8] Mizumoto M. *"Pictorial Representation of Fuzzy Connectives; Part1:Cases of t-norms,T-conorms and Averaging Operators"* IJFSS, vol1, pp193-214,1989.

[NN9] -Rast M. *"Application of Fuzzy Neural Networks on Financial Problems'*, Proceed. of the NAFIPS Conference,Syracuse,N.Y.,pp.247-349,1997.

[NN10] Fu L.Neural Networks in Computer Intelligence, Mc Graw-Hill, N.Y.1994.

[NN11] Snow P."*The Reasonableness of Necessity*" IJAR, 13, pp.287-301.

[NN12] Prade H."*A Computational Approach to Approximate and Plausible Reasoning with Applications to Expert Systems*" IEEE Transactions on Pattern Recognition and Machine Intelligence, 7, pp.260-283, 1985.

I. [AP] APPLICATIONS:

For many years, skepticism has reigned hampering the research among some scientific communities. As success of the application of fuzzy logic demonstrated in Japan, Europe, and elsewhere, the American Engineering community, realized that patents were awarded to foreigners, when the gauntlet had been discovered right here at the University of California, Berkeley. The few works listed here cover locolmotive applications, medical, computer vision, pattern recognition, financial, decision making, machine intelligence, expert systems, data-bases, and the way of the future, namely, Soft Computing, that means "computing with word rather than numbers". Last, but not least, the greatest, and most famous of its applications is that to control, so much so that it is debatable whether the use of ordinary differential equations is now to be considered obsolete. Each of the following works is in itself an excellent source of other remarkable achievements. For example, note the last entry by J. Buckley that examines the use of constrained fuzzy arithmetic in discrete probability calculations. So,again a wish for happy reading!

[**AP1**] Bonissone P."*An Expert System for Locomotive Repairs*" <u>Human System Managem</u>. 4, pp.355-462,1984.

[**AP2**] Adlassnig K.P. "*Approaches to Computer Assisted Diagnosis*" <u>Compu. Biolog. Med.</u>, vol 15. 1985.

[**AP3**] Keller J. "*The Impact of Fuzzy Sets to Computer Vision:*", pp.1-10., <u>NAFIPS Proceedings.</u> 1993.

[**AP4**] Jain A.K.& Backer E."*A Clustering Performance Measure Based on Fuzzy set Decomposition*" <u>IEEE Transact.on Pattern analysis and Machine Intelligence,PAMI-4</u>, 4, pp.357-363,1981.

[**AP5**] —Buckley J.J.,"*The Fuzzy Mathematics of Finance*", <u>IJFSS</u>,21,3, pp.257-273,1987.

[**AP6**] Schmuller J. "*Three Faces of Fuzziness Theory, Theory, Practice, and Applications*' <u>PCAL</u>,7, 2, pp.14-15, 1993.

[**AP7**] Mc Allister L.M.N. &Cebulka K.."*Transmittance Matrices,and an Evaluation Technique with an Appli cation*",in <u>Proceed.of the 3rd Conf..on Modeling and Simulation, University of South California</u>, Los Angeles, CA, pp.29-31,1981.

[**AP7**] — Mc Allister L.M.N. &Cebulka K&Galvin J.."*Stochastic Networks: an Evaluation Technique and an Application*"

[AP<u>Proceedings of the 13th Conference.on Modeling.and Simulation,</u> University of Pittsburgh, PA, pp.775-776,1983.

[**AP8**] —Bortolan G.&Degani R."*A Review of Some Methods of Ranking Fuzzy Numbers*", <u>IJFSS</u>,15,1, pp.1-19,1985.

[**AP9**] McAllister L.M.N."*Fuzzy Intersection Graphs*" <u>InternJournal Computers andMathematics with Applications</u>

[**AP10**] — Mc Allister L.M.N."*Simulation of Control for Automated Manufacturing*",

Proceed.of the Conf on Artificial Intelligence and Advanced Computer Techniques, Long Beach, CA, pp.83-88,1985.

[AP11] —McAllister L.M.N."*Inexact Graphs:Their Role in some Expert Systems*",Proceed. of the Conf.on Applications of Artificial Intelligence, Denver,4-7 December, pp.504-509,1984.

[AP12] —Zadeh L.A."*Fuzzy Sets*" Info. and Control, 8, pp.338-353, 1965.

[AP13] Turksen I.B.&TianY.&Berg M. ' *"A Fuzzy Expert system for a Service Centre of Spare Parts*" Intern.Journal of Expert Systems and Applications,, vol5, pp.447-464,1992.

[AP14] Yen J.&Langari R. Industrial Applications of Fuzzy Logic and Intelligent Systems, IEEE Press, New York,1994;

[AP15] Zimmerman H.-J.&Zadeh L.A.&Gaines B.R. (Eds.) Fuzzy Sets and Decision Analysis, Noth-Holland,New York,1983.

[AP16] Zadeh L.A. &Fu K.S. &TanakaH.&Shimura M. (Eds.) Fuzzy Sets and their Application to Cognitive Decision Making Processes, Academic Press, N.Y.,1965.

[AP17] Zadeh L.A. "*A Computational Theory of Dispositions*" IEEE Trans.Syst. Man Cybern., 6, pp.754-76339-63, 1987.

[AP18] Zadeh L.A. &Fu K.S. &TanakaH.&Shimura M. (Eds.) Fuzzy Sets and their Application to Cognitive Decision Making Processes, Academic Press, N.Y.,1965.

[AP19] KellerJ.M.&Carpenter C."*Image Segmentation in the Presence of Uncertainty*", Intern.Journal of intelligent Systems, 5,pp.193-208,1993.

[AP20] Jain A.K.& Backer E."*A Clustering Performance Measure Based on Fuzzy set Decomposition*", IEEE Transact.on Pattern Analysis and Machine Intelligence,PAMI-4, 4, pp.357-363,1981.

[AP21] —Buckley J.J.,"*The Fuzzy Mathematics of Finance*", IJFSS, 21,3, pp.257-273,1987.

[AP22] —McAllister L.M.N."*The Quality of Connection in a Network*" in Analysis of Fuzzy Information, by J.C.Bezdek, CRC Press, chapter9, 1987.

[AP23] —Zemankova M.&Kandel A. Fuzzy Relational Data Bases:A Key to Expert Systems, Verlag,TÜVRheinland,Kõln.

[AP24] Wang P (ed.) Advances in Machine Intelligence and Soft Computing, vol. IV,1997.

[AP17] -Yamakawa T. "*Stabilization of the inverted pendulum by High -Speed Fuzzy Logic Controller Hardware System*" IJFSS, 32, 2, pages 161- 180, 1989.

[AP18] - Yamakawa T."*Control of the Inverted Pendulum by Fuzzy Logic*", IFSA Proceedings (M.Sugeno, editor), volume2, Tokyo,July1987.

[AP19] Zadeh LA "*The Concept of Linguistic Variable and its Application to Approximate Reasoning*"Info Sciences, 8, pp. 199-251,&pp.301-357&pp.43-80, 1975.

[AP20] Esogbue A.&Elder R.C. "*Dynamic Programming, Fuzzy Sets and the Modeling of R&D Management Control Systems*" IEEE Transac.on syst.,man, and Cybernetics,13,1, pp.18-29,1983.

[AP21] McAllister L.N."*Availability and Competition: A Model* ", Mathem. Modeling, 9, 5, pp.477-490,1987.

[AP22] McAllister L.M.N. "The *w-Reciprocity Rule*" IJFSS,vol.23, no.7, pp.381-385,1987.

[AP23] Buckley J.J. Fuzzy Probabilities:A New Approach and Applications, Physics-Verlag/SpringerVerlag, Heidelberg, Germany, 2003.

[AP24] A. G. Gluckman& A.Celmins" *Cost Effectiveness Analysis Using Fuzzy Set Theory* ", ARL-TR-317, pp. 1-21, December1993.

[AP25] Serrano&Vila&Delgado"*Using Fuzzy Relational Databases to Represent Agricultural and Environmental Information: An Example within the Scope of Olive Cultivation in Granada*" Mathware & Soft Computing, vol.8, no.3, pp.275-289, 2001.

[AP26] RossTFuzzy Logic with Engineering Applications,McGraw,1995.

[U] Dealing with UNCERTAINTY MODELING:

After several years of reign as the lone science capable of dealing with uncertainty, the scientific community faced many instances of failures or of simple cases where probability did not apply. Thus, the need for different approaches became evident, see for example references [30], [AP23], [U3]. Therefore, because of the failures and inadequacies of the previous theories, belief theory, possibility, plausibility, necessity theories, and the whole field of the mathematical theory of evidence began to be studied and examined anew. Fuzzy logic, found its place too [U15]. Contributions such as [U9] are worthy reading.

Note the works on possibility theory, necessity, belief [U7], [U1, U8], and their tie to fuzzy sets [U5, 16].

[U1] Schafer G. A Mathematical Theory of Evidence, Princeton University Press,1976.

[U2] Zadeh L.A. "*The Role of Fuzzy Logic in the Management of Uncertainty in Expert Systems*" IJFSS, 11, pp.199-227, 1983.

[U3] Yager R.R. Possibility theory. Wiley1993.

[U4] Wassermann L."*Comments on Schafer's Perspective on the Theory and Practice of Belief Functions* "IJAR, Vol.6, pp.467-3764,1992.

[U5] Zadeh L.A."*Fuzzy Sets as a Basis for a Theory of Possibility*", IJFSS, 1, pp.3-28, 1978.

[U6] Rubens T.J.&Vincke P. Preference Modeling, Springer Verlag, 1985.

[U7] Provan G. "*The Validity of Dempster-Schafer Belief Functions*" IJAR, 6, 3, pp.389-400,1992.

[U8] Schafer G. "*Rejoinders to Comments on the Theory and Practice of Belief Functions* "IJAR,6, 3, pp.425-445,1992.

[U9] Wood K.L.&Antonson E.R.&Beck J.L."*Representing Imprecision in Engineering Design: Comparing Fuzzy and Probability Calculus*", Research in Engineering Design, 1, pp.187-400,1990.

[U10] ZadehL.A. '*Fuzzy Sets as a Basis for the Theory of Possibility*" IJFSS, 1, pp.3-28,1978.

[U11] Yager R.R.&Filev D.P. Essentials of Fuzzy Modeling and Control, Wiley,1994.

[U12] Zadeh L.A. "*Fuzzy Logic, Neural Networks and Soft Computing*", Communic. of ACM,7, 3 pp.77-86, 1994.

[U13] Dubois D.7Prade H."*A Set Theoretic View of Belief Functions*" Intern.Journal of General.Systems, 12,pp.193-226,1986.

[U14] Krishnapuram R. &KellerJ.M. *'A Possibilistic Approach to Clustering"*, <u>IEEE Transactions on Fuzzy Systems,</u> 1, 12, pp.98-110,1993.

[U15] Sztandera L. & KellerJ. M,."*Spatial Relations among Fuzzy Sets of an Image"*, <u>PROCEED.ISUMA</u>, Univ.of MD, College Park,pp.207-211,1990.

[U16] —Zadeh L.A.& Kacprzyk J. <u>Fuzzy Logic for the Management</u> <u>of Uncertainty</u>, Wiley&Sons, NewYork,1992.

[U17] Zadeh L.A. *'Fuzzy Sets as a Basis for the Theory of Possibility"* <u>IJFSS</u>, 1, pp.3-28,1978.

[U 18] Pearl J. <u>Probabilistic Reasoning in Intelligent Systems:Networks of Plausible</u>, Morgan Kaufmann, Morgan Kauffmann, 1988.

[U19] Schafer G. <u>Probabilistic Expert Systems</u>, SIAM, Philadelphia,1996.

PART II

APPLICATIONS

CHAPTER 9: "SOLUTION OF THE INVERTED PENDULUM STABILIZATION BY FUZZY LOGIC METHOD"

In this section, we plan to discuss a few reflections that are a consequence to the use of fuzzy logic in applications particularly to control Over the past years, we grew accustomed to teach ordinary differential equations as the only method to handle control problems as they occur in industrial situations. Fuzzy logic has introduced an additional tool that has proven to be fairly simple to use and highly successful. We report here a few basic concepts, and the example of its use for the problem of the inverted pendulum.

SECTION 1. BACKGROUND ON CONTROL SYSTEMS.

Assuming that few readers are familiar with the pertinent terminology, we begin with a review following T. Ross [13, pages 469-472]

DEFINITION:

A control system is an arrangement of physical components designed to alter, regulate, and command through a control action another physical system behavior.

Control systems are typically of two types:

(I) **OPEN loop** systems, In which case the control action is independent of the physical

Examples of this type are a toaster, an automatic washing machine.

(II) **CLOSED loop**, systems, also called feedback systems because the control action depends on the physical system output.

Examples of this type are the room temperature thermometer to control the heating or cooling of an environment.

The two systems that have mentioned are respectively called *"sensor"*, for example a thermometer or an optical device, if it is the system that controls; the other system is called the *"plant"* since it is the system that is controlled.

Generally, in a closed loop case, certain signals of the system, also called inputs, are determined by the response of the system, also called outputs. see Figure 1 below.

FIGURE 1. A typical configuration of a closed-loop system. [13, page 471] where S stands for sensor component; P for plant; C for Controller, or Compensator; and Σ for system input, Ω for system output that will return operation to the sensor and then back to the controller

$$\Sigma \rightarrow C \rightarrow P \rightarrow \Omega$$
$$\uparrow \leftarrow S \leftarrow \downarrow \lrcorner$$

As a quick introduction, leaving,however most details to section 9, we mention that in the case of the inverted pendulum, where the goal is to move a vehicle and keep a rod in a position as close to the vertical as possible. thus, a small motion will cause an increase, in a direction or other, of the angle of the rod with the vertical, thus a corrective motion of the vehicle is needed to rectify the position of the rod, and so on. We have a closed loop with a system input, sensor, controller operating according to a given number of linguistic rules as specified by the matrix of rules, see later pages 45 to 53, or better yet [3, pages 340 on], with the variables possible valuations listed on page18 according to (4) and (4a).

FIGURE 1. The system input, whose simple scheme consists of the vehicle that is denoted by V, next page, that can move left or right on wheels, the rod, namely the vertical that is drawn dotted to make it clearly distinct from the rod making a possible angle that is denoted e in the figure next page on figure 2, on page 53 whose rate of change will be denoted De.

FIGURE 2.

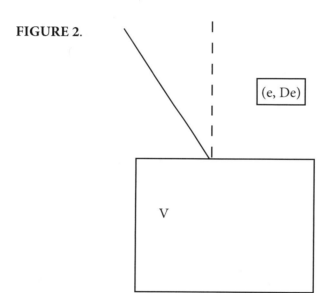

Comments on figure 2. V indicates the Vehicle possibly on wheels; e is the angle that is formed by the position of the rod (drawn with a solid line to distinguish it from the vertical) with the vertical; De is its rate of change; both are bits of information {gathered by an optical sensor} fed to the fuzzy logic controller that will then decide according to its knowledge base on the motion w of the vehicle to move it the left or right to stabilize the ro

FIGURE 4. [3, page 341]. The possible choices of valuations for the angle e are listed along the top row and they are chosen from the lists on page 18 (4) or (4a). Those of its rate of change De are listed along the first column. The result valuation for the motion w of the vehicle V of a rule is found reading the matrix as usual. For example, the rule *"If e is NS and De is NS then w is PS"*. Likewise we have the rule "*If e is PS and De is NS then w is AZ"and "if e is AZ and De is AZ then w is AZ too*"

De \ e →	NM	NS	AZ	PS	PM
↓					
NS		NS		AZ	
AZ	NM		AZ	PM	
PS		AZ	PS		

The above figure basically shows the rules that form the base for the decision making capability of a fuzzy controller.

FIGURE 5. The system of differential equations that we ordinarily study for the inverted pendulum is thus as follow as in [Klir &Yuan, see [2, p.340]:

IDDe =VI sine + HLcose;

L-mg = mL{DDe sine + [De] ^{2} } cose

H = mDDw + L (De cose - De] ^{2} sine).

For clarity of the comparison discussion, we assume again little or no recall of this experiment, thus we have reviewed the system of differential equations, see below figure5., and the physical configuration of this case, see the simplified scheme of figure2. where the following variables are used and we repeat the significance of its variables, although it was done so already on the previous page.

1) - The angle between the rod and the vertical as in figure 2.denoted e;

2) - Its first and second derivatives denoted respectively De, e dot, and e double dot or DDe;

3) - The length of the pendulum denoted 2L;

4) - The second derivative DDw, or w double dot, of the position of the vehicle denoted w;

5) - Finally, m and M denote respectively the mass of the pendulum, and of the vehicle;

6) - In addition, we have H and V denote the horizontal and vertical forces of the pivot where by V we mean the driving force given to the vehicle; and I is the moment of inertia, therefore I = (L ^{2}) /3.

Following the review, that is generally presented, as in [2, pp 340] we have the system as we reported above.

SECTION 2. THE FUZZY CASE

We shall call *"fuzzified "*the version we obtain by using fuzzy logic methods. In the *fuzz-*

ified case, there is an additional sensor, which has the purpose of monitoring the process of operations much like a human operator overseeing the process would. In contrast, we will use the term *crisp* to denote the ordinary case where the notion of belonging, or of truth, is clear there is no difficulty or ambiguity in making a decision of what it is. Thus, the latter sensor will bring about modifications according to its knowledge base. We must remark now that at the core of the fuzzy knowledge base are linguistic rules that are expressed by fuzzy statements namely statements of the type

$$\text{"v is F".} \dots \dots \dots \dots \dots \dots \dots \dots \dots \dots \dots \dots \text{(1)}$$

where v is a variable of the universe of discourse and F is a fuzzy attribute, or a fuzzy set, namely a set with undefined boundaries so that we must specify the degree of belonging of an element by attaching a membership function for each element to eliminate any possible ambiguity.

The *"modus ponens "* in fuzzy logic is much like the classical case as we shall illustrate next.

SECTION 3. SOME MORE DETAILS.

How does it all fit together? The expression (1) is a fuzzy proposition. Note that in sharp contrast with classical logic where there exist only two quantifiers, in fuzzy logic, we have several such as any one from the following list:

$$\{\text{Several, often, small, large, about, approximately}\}. \text{ (2).}$$

It is safe to say that all of the above quantifiers express some degree of approximation, not to be confused with probability. Of course, other quantifier may be added to the list (2) above. As a matter of fact we shall do so in the list (3) as it apples to the case of the inverted pendulum.A review of sections 1 to 8 of part I. may be a good idea.

S4. HISTORICAL PERSPECTIVE.

The international Fuzzy Sy stems Association, consisting of several thousands members from Europe, Asia, North and South America, holds annual meetings. In July1987, the meeting was held in Tokyo, Japan. The meeting included besides the proceedings publishing the text of a selection of contributed papers, the keynote address by Professor L.A. Zadeh, an exhibit of results by the fuzzy engineering community. Some were software packages as demonstrated by J.Buckley by M.Sugeno. Others, as that by T.Yamakawa, provided a video demonstrating the solution for the control of the inverted pendulum. This fundamental paper was included in the proceedings and was published again, [5].

SECTION 4. THE FUZZY AND THE ORDINARY DIFFERENTIAL MODELS.

In TABLE 1,on page 58, we copy the variables and the system of differential equation as stated in Reference [3, page 340] that matches the description in most mathematical physics texts. The **goal** is to move a vehicle so that the inverted pendulum remains in a vertical position. The motion is controlled by the F.C. For clarity of the discussion, we assume again little or no recall of this case, thus we review the system of differential equations and the physical configuration of this case respectively figure 3 and FIGURE 2, on page 49, where the representation of the physical model helps us understand the variables and the linguistic rules as they are then stated in the following figure. Both Ross [13] and Klir&Bo [3] texts do an excellent job in presenting this application and others. Thus, we shall follow their lead, making sure that Yamakawa's original model holds. Recall as we already discussed previously in Figure 1, on page 46, and on figure 2 on page 51, the system input will be connected to a sensor and to the controller that will feedback to the vehicle by correcting the motion to the left or to the right, just as a human operator would if observing, or as if playing a game.

It is an appropriate time to mention now why control by fuzzy logic has become so successfully implemented in manufacturing operations [MOF11]. It requires only to learn the rules of a human operator, translate them in a suitable fashion to the fuzzy controller, and there you have it! all done !……….. It is cheap, it is simple. What else do we need? See the section on popular press, [PP] in the references as well.

TABLE 1: As already reviewed, and repeated for clarity, in the methods for example using ordinary differential equations, the variables are:

(1) - The angle is denoted e;

(2) - Its first and second derivatives are denoted respectively D e and DDe;

3) - The length of the pendulum denoted 2L;

4) - The position of the vehicle V is denoted w and its second derivative is denoted DDw;

5) - Finally, m and M denote respectively

[3, page340] The ordinary differential equations as needed for the stabilization of the inverted pendulum, where H, and V equal respectively the horizontal and vertical forces:

I= (L^2) / 3.

I DDe=VL (sine) - HL (cose)

V – mg = mL (DDe) (sine) + (De) ^{2} (cose)

U – H = M (DDw)

where the symbol D indicates the first derivative with respect to time, and DD indicates the second derivative.

Although the meaning is noted for each figure, we prefer to risk repetitiousness for sake of clarity the mass of the pendulum, and of the vehicle;

6) - In addition, we have Hand V that denote the horizontal and vertical forces of the pivot where by V we now mean the driving force given to the vehicle; and I is the moment of inertia, therefore I = (L ^{2}) /3.

Although the meaning is noted for each figure, we prefer to risk repetitiousness for sake of clarity

Presumably, the configuration must also include a sensor, an optical device, a fuzzy logic unit, all arranged as in a closed loop, according to the indicated flow in the review of the closed loop control method, namely in figure 1 on page

The needed number of linguistic rules, see section 5, is not an easy decision to make. Are there any guiding rules? This is a topic that is still very much under research. Up to now this number varied with the task on hand.

N.B. For the next two figures we use: N for negative, P for positive; S for small; M for medium, L for large; A for approximately; Z for zero"

CONCLUSIONS

Are Differential Equations Becoming Obsolete? Should we not include the teaching of fuzzy logic within the undergraduate curriculum? To conclude this simple overview. What is then fuzzy logic? In a nutshell, it is a mathematical framework that much like the human brain is able of making sense out of inexact information such as garbled speech, or indistinct pictures. Is there a theory backing its development? Yes, it is fuzzy sets theory, see e.g. [3] or [4] that has been introduced since 1965.Unfortunately, the fact that was largely ignored by the mathematical community, is unexplainable, Yet, it has been explored abroad, see the section on [MOF] on page 34 that includes excitingly beautiful contributions. However, it was soon detected by the engineering community that is now trying to catch up so that we do not continue to have most USA patents awarded to foreign scientists when the gauntlet was introduced by one of ours, professor L.A. Zadeh, director of BISC, of the University of California, Berkeley. Soft Computing is also prospering abroad, just to name a few, there are centers at the University of Barcelona, Spain, and at the Applied Mathematics Seminar at the University of Gent, Belgium, at the school of operations research in Aachen, Germany, and here at the University of California, Berkeley, at Texas A &M. Finally check the texts [AP18], and the whole section [AP] listed in the references is an eye-opener for those who still hesitate to take this subject seriously.

A common question is how to construct the fuzzy logic controller. Success has been achieved in this matter by several. Most notably, Takeshi Yamakawa, dean of the engineering school of Kyushu University, Iizuka, Japan has a microchip available for purchase and several papers that describe its construction.

APPENDIX

Suggested exercises and a summary of Basic Concepts that form the basis of fuzzy set theory for fuzzy logic to be discussed:

Even though this part is not strictly necessary, this appendix is an addendum to this text because its purpose is that of providing the mathematical background for fuzzy logic. It turns out that it may provide a source of projects for the enterprising reader. Some of the exercises have answers provided on page 4.

We begin with a formal definition and a question of interpretation that should follow easily from the remarks on the distinction between two concepts that if not well understood, it causes much confusion.

DEFINITION 1: A fuzzy set is the set of pairs $\{ (x, m(x) \}$

where x belongs to some universe U and m (x) is a continuous function,.

Note: For some authors, the support set of the fuzzy set needs to be bounded as well!

Question#1: Explain the distinction between fuzziness and randomness; Express it in sentences or give at least three different examples.

DEFINITION 2: the **Cardinality** of a fuzzy set is defined as the sum of the membership grade of all the elements of the set.

Question #2: Let A and B be two sets defined in the same universe U. Prove that $|A| + |B| = |A \cup B| + |A \cap B|$;

Question #3: For each of the following membership function:

♦ a (x) $= x/ (x+2)$;

(ii) B (x) $= 2^{-x}$;

(iii) c (x) $= 1/ [1+10 (x-2)^2]$;

Find the membership expressions of their negation operators;

NOTE: Originally,Zadeh formulated a fuzzy set as a finite number of pairs thus we say that we have the discrete case as in A (X) = [.4/P+.5/q+.5/r+.4/s=1/t} where the pairs are denoted by.4/p instead of (.p,.4); Let the second set be a continuous case such as for B (x) $= \{x/ (x+1)$ for $x \varepsilon [0, 10]$

Find their cardinality.

CHAPTER 10: FUZZY RELATIONS

In this chapter, our goal is to present how we derive the *fizzification* of classical concepts such as that of relation and its representation, namely a graph.

DEFINITION A *Fuzzy Relation* is a set of pairs (a,b) from an ordinary cross product of A and B associated with a continuous, or a discrete, membership function denoted m (a,b) usually with values from the unit interval. It will be denoted, using again f to remind that we fuzzified an ordinary definition.

R (f) = { (a,b), m (a,b)) /a e A, b e B/ m (a,b) e [0,1] }.

where by A and B we denoted any two classical sets that will be called crisp to juxtapose the ordinary concept of set to that of fuzzy set that plays a key role in fuzzy logic.

When an element is depicted by a point that is called either **vertex** or **node**, and the pair (a,b) is depicted by the **edge**, also called **link**, joining the vertex a to the vertex b, we have what is called the pictorial representation of a relation. Likewise we have the equivalent, namely we have a

A FUZZY GRAPH

More specifically, in the fuzzified case, we have if

R (f) = { (a,b), m (a,b)) /a e A,b e B, m (a,b) e [0,1] }

is a fuzzy relation then G (f) is its depiction, thus, a **Fuzzy Graph** was originally defined as the graphical representation of a fuzzy relation [MOF15]. Again, we must realize though that it is a pair of fuzzy sets, the fuzzy vertex set and the fuzzy edge set [[4, AP11,AP22,AP23.AP10,MOF14].

NOTE:

Again we often hear objections to the membership function which a few stubbornly insist that it is a probability. The only thing it has in common is that its valuation is from the unit interval. However, the meaning is strictly semantic related as we emphasize in the next example, and in many others, that can be easily constructed where the valuations have totally distinct meaning.

EXAMPLE. We may say "*The probability that Kevin and Kristin date is* 0.9" but if we say "*Kevin's commitment to Kristin is 0,9*", we conclude the valuation in the latter case is absolutely different from the meaning of the valuation in the former case.

PRACTICE QUESTIONS

Question #1-**part (I)** - Give an example of semantic for a pair associated with a probability value; and then an example of semantic for a pair that has a membership grade; **part (II)** -Let a fuzzy binary relation R be defined on the set A={x/x is an integer between 1 and 100} and B ={Y/y is an integer between50 and 100}

According to the statement "x is much smaller than y" with membership

R (x,y) = = 1-x/y, if x<<y;

=0 otherwise;

QUESTION #2. Answer each of the following: (I) -What is the support set of R? **(II).** What is the range of R? **(III)** Is there an inverse?

QUESTION #3. Determine the expressions and the sketches of the membership functions of the following fuzzy sets: a) A^*, B^*, C^*; b) $A^* \cap B^*$; c) $A^* \cap C^*$; d) $B^* \cap C^*$;

Where *denotes the fuzzy complement operation.

DEFINITION 4: A *fuzzy Number* is a **convex** fuzzy set.

To answer the frequent question:

"**How do we generate fuzzy numbers?**"

We present an algorithm as proposed by A.Kaufman, the author of one of the earliest books on fuzzy set theory;. His **Algorithm** interestingly produces what can be called the equivalent of fuzzy integers.It roughly consists of the following steps:

Step 1- Choose a seed function (possibly one that satisfies the condition of a fuzzy number) that will be denoted f_1 (x);

Step 2- Derive f_2 (x) as the definite integral over [0, x] of the product of f 1 (x) with f_1 (x −t);

Step 3-Continue the process with f n (x) equal again the definite integral over [0, x] of the product of $f_{(n-1)}$ (x) with f_1 (x −t);

Step 4- Look for a pattern to show so that you can find an explicit expression for f_n (x);

More Practice Questions:

QUESTION #5-Use the above procedure for each of the following seeds:

f_1 (x) a e $^{-ax}$, a>0; (ii);

QUESTION #6-Determine which of the following sets are fuzzy numbers. Justify your answer.:

a) -A (x) = = sin x for x ε [0,π];

=0 otherwise.

b) -B (x) = =x if 0≤x≤1;

=0 otherwise.

c) - C (x) = =1 for 0<x≤10;

=0 otherwise;

d) - D (x) = Min (1,x) if x≥0;

0, if x<0;

e) -E (x) = =1, if x=5;

= 0, otherwise.

More answers are provided on page 67 and following.

DEFINITION 5: The complement of a fuzzy set with membership grade f (x) is the set with membership grade

g (x) =1 – f (x).

In which case, we refer to it as simply the **Negation Operator**. There are at.

(least two well-known definitions for complements. They are:

(I) Yager's [3, page 57]: Where the membership grade for the fuzzy complement is given by

$$Y_N (x) = (1 - x^N)^{(1/N)}, \qquad \text{with } 0 < N < \infty$$

(II) Sugeno's [3, page 54]: In this case we have:

$$S (x) = (1 - x) (1 + ßx), \text{ with: } -1 \leq ß < \infty$$

Since both are function of either N or of β, we refer to them as classes. Use a graphing calculator, and sketch several curves cirresponding to a few different values for N and for b.

(III) McAllister [AP22] noted that it is sometimes necessary to soften the negation. For example, we might say that we do not like cherries because we prefer another fruit, not because we dislike them

(III) There has been a lot of interest in the use of fuzzy relations in decision making [1, 2, 3, 4] on page 49.

A twist is discussed next, [5].

0. INTRODUCTION:

Given a small number of choices, a matrix is a convenient way to represent a fuzzy preference relation. Usually, the matrix is reciprocal. The **W-RECIPROCITY RULE** that is presented here as a way to relax the rigidity of the reciprocal assumption and to embody linguistic meanings for indifference and for the representation of the converse preference.

1. PRELIMINARY REMARKS

Let the matrix

$$R= (r_{ij}), i,j = 1,…,n ……………… (1.1)$$

Describe a preference relation. The assumption that for all i and j, the matrix satisfies the reciprocal rule means that its entries satisfy the condition

$$r_{ij} + r_{ji} = 1 ……………….. (1.2)$$

if j ≠j, otherwise r_{ij} =0, [1,2,3,3,4], see figure1, where r_{ji} denotes the converse preference.

W-FIGURE1. The ordinary reciprocity rule is sketched by a line, say l, of equation y =x. Its converse by a line, say l_c, of equation y =1-x, as we see below. These two lines meet when x=1-x, or 1=2x, or x=0.5. If the meaning of x=1 is complete preference, y=0 is dislike, then y=0.5 may mean indifference. It is suggested that the reader sketches the two lines that are mentioned above and, that will easily fit in the space below.

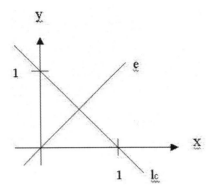

This way, if the value of the preference for choice i over choice j denoted r_{ij} is known then the value of the converse preference, denoted r_{ji} is simply computed from (1.2).It is a handy assumption as the next example points out.

EXAMPLE 1. The interior and exterior of a company car must differ. The colors are chosen by a committee of 12 individuals, and the options are [yellow, green, red, black}. In which case, there are 12 matrices, each of dimension four by four. Besides time consuming, the procedure of expressing the value of the preference for one color over another is confusing and boring. When the reciprocal rule holds, instead of entering $(16 - 4) \times 12 = 144$ data only a total of 72 input data is needed.

I the special case can be easily generalized. Namely, given n choices and m individuals, the reciprocal rule allows the total number of input data to be reduced to a total of

$$(1/2) m (n^2 - n)$$

input data. It seems that no other remarks are necessary. However, the reciprocity condition is too rigid and not realistic, as the next example points out.

EXAMPLE 2. Let the choices in fruits that are available be {cherries, seedless grapes, and bananas}. In the following, we shall use the first letter to denote a type of fruit, rather the subscripts. This way,

$$r_{gc} = 0.7$$

denotes the preference of seedless of the preference are clustered about 0.5, thereby often denoting indifference, it is reasonable to expect that such indifference holds even more so for the converse. In such a case, there ought to be a band of indifference about the point 0.5, see figure 2.grapes over cherries. However, to derive that

$$r_{cg} = 0.3$$

may be not true. It may in fact have a far less value because cherries have pits.

Can we release, and replace the rule (1, 2) with a less rigid and more realistic condition? It is the goal of the discussion in the following section.

2.W-RECIPROCITY RULE

The desirable features of a relaxation of condition (1, 2) results from the recognition of its shortcomings. First of all, when the values of the preference are clustered around 0.5, we understand that there is indifference; in which case, there ought to be a band, for example of width denoted 2w, of indifference about the point 0.5, see the W- Figure 2 on the next page after the references.

REFERENCES:

for this section, on the reciprocity rule:

[1] —Bezdek J,Spillman B.,Spillman R.*A Fuzzy RelationSpace for Group Decision Theory*, IJFSS,1,pages255-268.1978.

[2] —Blin J.. M.Fuzzy *Relation in Group Decision Theory*", Journal Cybernetics, 4, pp.17-22,1974.

[3] —Kacprzyk J.'*Group Decision Making with Fuzzy Linguistic Majority*,IJFSS,18, pp.105-118,1986.

[4] Nurmi H. "*Approaches to Collective Decisionmaking with Fuzzy Preference Relations*,IJFSS,6,pp.249-259,1981.

[5] —Mc Allister L.M.N. "*The W- Rreciprocity Rule*', IJFSS, 23, 4, pp.381-385,1987.

W-FIGURE 2. Since x=0.5 has the meaning of indifference, consider the interval [0.5-w. 0.5+w]. To each point within this interval we assign the value 1. Consider the points (0,0) and (1.0) on the x-axes that we wish to connect to the end points of the indifference band we created over the interval of width equal to 2w about x=0.5.

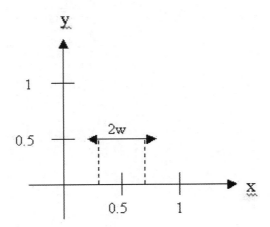

Sometimes, we wish to include some linguistic features as well. For example, it may happen that the converse preference is actually *slightly higher* than $1 - r_{ij}$ conveys. Most importantly, as the values of the preference cluster near 1, namely when the expressed preference is clearly strong, it is expected that the preference for the converse may very well preserve the characteristic of strength. In this case, the situation is described by a *slightly higher* value, and the generalization is said to be

EXAMPLE 3. Suppose we are given the choice of out of three colors

{Maroon, blue, gray }

Assume we express a preference by stating that

"Maroon is much better than blue or gray".

Should maroon be no longer available then we might say

"Blue or gray are acceptable". Using again the first initials of the colors, and let the quantification be

r_{mb} =0.9, r_{mg} = 0.8; r_{bg} = 0.6. Under the (1,2) rule then the converse preferences are given respectively by 0.1 and 0.2. Neither takes under consideration that blue is always mentioned first in each statement. Can we include that the indifference between blue and gray remains about 0.5?

Consider the interval [0.5- w, 0.5+w] Let our attention be directed to the construction of a function W (x, w) that supplies the strength preserving case. How do we wish to modify the rigidity of the previous rule? Rather than a triangular region we may create a trapezoidal region that includes the indifference band as one of its sides.

We suggest the creation of a function that will be denoted W (x, w). Thus we shall proceed as we describe in the following: **(I)** First, we subdivide the unit interval into three parts, namely:

[0, 0.5-w], [0.5-w, 0.5+w], [0.5+w, 1], so that on the second subinterval indifference holds-, i.e. W=0.5, for all values of x;

Then: **(II)** We define a function denoted W (x, w) that will join the indifference band at its endpoints on the second subinterval;

So that it is concave up on the first, and third subintervals;

For example, let

$$W (x, w) = (1 - x) / (1 + ax).$$

if x ε [0, 0.5- w];

$$W (x, w) = (1 - x) / (1 + bx)$$

if x ε [0.5 +w, 1];...... .... (2.1)

How do we find the evaluation s for a, and for b? We use the fact that we want it to join at the endpoints as we re -quested in (II), see next in (2.2)!

$$W (x, w) =$$

0.5 otherwise;

Note that if w=0 and a=b=0 then the sketch of the function W reduces to a straight line yielding the reciprocal rule. We thus conclude that the function W is a generalization of the reciprocal rule. Furthermore we have:

W (0,w) =1 and W (1, w) =0,

providing us with a continuous function.

Thus, at the end points, of each subinterval, we have:

a = 4w/ (1 – 2w), b = -4w/ (1 + w), if w ε [0, 0.5] .... (2.2)

to obtain the sketch that completes, as desired, that in figure 2.

W-FIGURE3. W-RECIPROCITY FORMULA AND ITS APPROXIMATE SKETCH:

(1 –x) / (1 +ax) if x ε [0, 0.5- w]

W (x,w) = { (1 – x) / (1 +bx) if x ε [0.5 +w,1] (2.1)

0.5 otherwise.

Where we also have:

$a = 4w/ (1 - 2w)$, $b = -4w/ (1 + w)$, if $w \varepsilon [0, 0.5]$.

Note that this function W (x, w) creates two concave up branches joining the points $(0,0)$,and $(0.5, 0)$ respectively to $(0.5-w, 1)$ and $(0.5+w, 1)$, creating the approximate sketch below.

branches joining the points $(0,0)$,and $(0.5, 0)$ respectively to $(0.5-w, 1)$ and $(0.5+w, 1)$, creating the approximate sketch below that is completed and corrected by a <u>descending</u> concave-up curve joining $(0,1)$ to $(0,5-w, 0.5)$

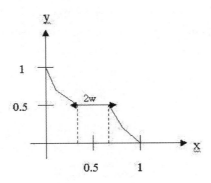

FIGURE. The sketch corresponding to the more relaxed rule to find the converse.

To conclude, the values of the converse preference will be derived from the W function.

EXAMPLE 5. Assume that $r_{12} = 0.6$; $r_{13} = 0.4$; $r_{14} = 0$, $e_{23} = 0.7$; and $r_{34} = 0.8$.

With an indifference band of $w=0$, the intervals are $[0, 0.4]$, $[0.6,1]$, and $a=0.5$ $b = -0.3333$.

Then we have $r_{21} = r_{31} = r_{32} = 0.5$; $r_{41} = 1$;

$R_{32} = 0.391$; $r_{42} = 0.857$, $r_{43} = 0.272$. Changing the indifference band value and repeating the same computation as in the above, we verify the results listed in the two matrices W $(x,0.1)$, W $(x, 0.2)$ below.

w=0.1   w=0.2

0	0.6	0.4	0	0	0.6	0.4	0
0.5	0	0.7	0.1	0.5	0	0.7	0.1
1	0.857	0.272	0	1	0.794	0.368	0
0.5	0.391	0	0.8	0.5	0	0.368	0

As expected, the w –preference scheme is sensitive to the magnitude of the indifference band. For the *strength preserving* case, the larger the band the smaller the difference between the preference of the i-th choice over the j-th choice and its converse. The calculation of the last example confirm what we expected because in the example, the preference for the third color over the fourth is 0.8. if w=0, the preference for the converse is 0.e; if w= 0.1 the converse equals 0.272; finally if w= 0.2 the converse equals 0.368. Only minor modifications are necessary should the *strength reversing* be desired.

It remains to illustrate in what way; the magnitude of the band w is selected.

There are already several methods that have been proposed over the years.

However, an additional simple way is discussed next.

Recall that one of the purposes of w is to indicate the magnitude of the indifference, which transpires in the responses of an individual. Thus, it ought to reflect how many values of the responses cluster about 0.5. For each individual k, with k = 1,….., m, determine an initial rating of the choices with the agreement that (1) no value needs to equal the lowest value 0 nor the highest value 1, and (2) any value may be repeated. Let d_k equal the span between the highest and lowest value respectively different from 1 and 0. Let p_k equal the number of values in the sequence which are less or equal; to d_k. Then we have

$$2w_k \leq d_k / p_k, \ k = 1, \ldots., m$$
$$w = \min \{w_k\} \text{ for } 1 \leq k \leq m \ldots\ldots\ldots\ldots\ldots\ldots\ldots\ldots\ldots\ldots\ldots (2.3)$$

EXAMPLE 6. Given the four colors choices of example2., and two individuals' ratings (0.6,0.6,0.9,0.2) and (0.3, 0.8, 0.8, 0.4) then we have according to the scheme above

$w_1 = 0.117$; $w_2 = 0.2$; $w = 0.117$. To conclude again, w, the w-reciprocity scheme (2,1) (2,2), (2,3) is simple to use and it adapts readily to situations which should not be described simply by the reciprocal rule (1,2).

QUESTION #6: part (I) - Plot on a graphing calculator both complements membership functions with w = 0.5,1,2,5 for Yager's; with ß =10,2.0.-0.5.-0.9 for Sugeno's.

   Note that for N=1and for $\beta = 0$ we have the simple negation operator.

   Part (II) compare the two sketches, are there any conclusions?

DEFINITION 6-If we define as the *"point of equilibrium"* the point s for which it is c (s) =s, namely where there exists a fixed point; See question 9 below.

QUESTION #7 - Prove that every fuzzy complement function can have at most one equilibrium point.

   Requirements for any fuzzy complement function are:
   1) - c (s) is monotonic;, namely if u>v then c (u) >c (v);
   2) c (2) =c (0) =1, c (1) =0;
   3) c (x) is involutive, namely we have c (c (x)) =x.

QUESTION #8. Is there any reason for those requirements? Check the first two requirements whether they hold if given any m (x), then the negation is

   c (x) =1-m (x).

QUESTION #9. Find the point of equilibrium, namely when f (x) = x, for Yager's and for Sugeno's functions.

QUESTION #10. If the empty fuzzy set is defined as the set with membership grade m (x)

=0, prove that for any fuzzy set we have: $A \cap \emptyset = \emptyset$; If $A \cap N(A) = \emptyset$, where N denotes the negation operator, prove that $A \cap N(A) = \emptyset$.

QUESTION #11-Show that the De Morgan Laws hold for fuzzy sets.

QUESTION #12. Find an expression for m (a) for figure 4 in section 6.

QUESTION #13. Show that both the negation operator g (x), and Yager'sY_N (x) and Sugeno's $S_ß$ (x) complements are involutive.

Some answers to the exercises:

#8. c (m (x)) =1- [1-m (x) =1-1+m (x) = m (x). Therefore involution holds.

Next:
c (0) =1-0=1, and c (1) =1-1=0. Unfortunately, it is not monotonic because If u< v, -u>-v, then 1-u> 1-v., or c (u) > c (v).

#1. Randomness refers to an event that may or may not occur; fuzziness refers to the boundary of a set that cannot be made precise.

For example, we have random car accidents referring to the probability of one to happen; fuzzy set if we refer to the severity of accident..

#2-Cardinality;if S (f) = {x/0.1,y/0.3+z/0.3+u/0/5+w/0.6} where the symbol/denotes association, then |S (f) | = 0.1+0.3+0.4+0.5+0.0.6=1.9;

Proof: $\Sigma_{x \cup}$ |min A (x), B (x)] +max [A (x),B (x)] =: $\Sigma_x \varepsilon_U$ |A (x),B (x) |=
: $\Sigma_x \varepsilon_U$ |A (X) + $\Sigma_x \varepsilon_U$ [B (x) |= |A|+|B|.
#3. (i) -1 - [x/ (x=3)] =2/ (x+ 2);
(ii).-: 1 - 2^{-x} = 2^{-x} x -1] / 2^{-x}
(iii)
ans. =$^{(x-2)2}$[1-10 (x - 2)^2]; and their sketches are:

CHAPTER 11: ARE THERE APPLICATIONS OF FUZZY GRAPH?

Yes, there exists the following paper which is reproduced entirely, including its references on page 84.

1. BACKGROUND.

The original definition of a fuzzy graph was that it simply provided us with a pictorial representation of a fuzzy relation. What is the original definition of a fuzzy graph?

Given two ordinary set A and B, we form an ordinary relation by considering their cross product, [10].

If to each element (a, b) in A x B, we associate a membership function m (a, b), then we form the fuzzy set

$$(\{ (a, b); 0 \leq m (a, b) \leq 1\}\ldots (1)$$

Some, realizing that since the values of m (a, b) are from the unit interval, immediately claimed that the membership was a probability value.

This way, the misunderstanding escalated into a controversy that is promptly resolved after we note that the difference between membership grade and the probability concept is quite great.

Basically the difference is explained by an example emphasizing the distinction in semantic.

The function m (a, b) quantifies the grade of membership of the pair to the fuzzy set. For example, suppose we *say < The probability that Matt and Jill are dating is 0.9 >*.

When we say *< The commitment of Matt to Jill is 0.9>* the latter numerical value portrays an assessment of the seriousness of the commitment between the two individuals, not the measure of the probability of their dating.

Thus, in both cases, we may associate to the pair (Matt, Jill) the same value 0.9. However, the significance of the resulting set is enormous.

Thus a fuzzy graph depicts the fuzzy set as in the notation (1) above, namely we have

$$G (f) = \{ (a, b); m (a, b) \text{ ¤ } [0, 1]) \}\ldots (2)$$

Where (f) reminds us that the resulting set is fuzzy.

3. A NEW DEFINITION OF FUZZY GRAPH.

The new definition states that G (f) as a pair of linear vector spaces. It remains to explain how it is possible to define linear independence for the edge set and for a vertex set of a graph.

Consider a widely accepted definition see Reference [2, chapter5]:

DEFINITION 1. we say that a **set of vertices V**, of *nodes* **is linearly independent** if no two vertices of V are joined by a link, or edge.

DEFINITION 2. A **set of edges E,** or of links of a graph is said to be **linearly independent** if no two links in the set have a vertex in common.

EXAMPLE 1. Let the set links E= { [1, 2], [1, 3], [1, 4], [2, 4], [1, 6], [6, 4], [5, 4], [3, 4] } where the vertices are shown in the figure 1, see also [2, p.175].

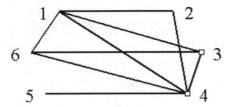

(FG) -Figure 1. The above figure shows a graph whose vertex set V={1, 2, 3, 4, 5, 6} has maximal independent sets, or a base, that by inspection we find to be {1, 5}, {2, 3, 5}, and {2, 5, 6}, {4}, see [2, page 175]. Note that the links [1,2], [6, 3], and [5, 4] have no vertex in common, thus they are linearly independent and form a base for the edge set E.

4. THE ADVANTAGE OF THE NEW DEFINITION

Why do we have a new definition of fuzzy graph? Its advantage is that if, for each vector space, once we can find a base, then any link, or vertex that fails, can be expressed as a linear combination of the elements of its base, or of its maximal spanning tree.

Is there a contradiction between the former definition of a fuzzy graph as a pair of labeled graphs and the current as a pair of vector spaces? The question has been answered in a negative way in [3].To be more specific, let

$$\{E_1, E_2, E_3, E_4, e_5, E_6, E_n\}$$

be a base we seek and found by the method in [5], or that in [6], we must be able to argue that if it forms a base of the support crisp link set E of the fuzzy set E (f) where again (f) denotes the fuzzy case then the corresponding set of labels we denote here as

$$\{L_1, L_2, L_3, L_4, L_5, L_6, L_n\}$$

form a base for E (f).

An argument by contradiction will yield the desired conclusion. Note that a similar argument holds for the fuzzy node set V (f).

5. HOW DO WE FIND THE BASE OF A GRAPH?

If the graph is relatively small, inspection will do. If large, we need other ways. Before mentioning two excellent techniques, we will discuss how it is possible to use a base to replace a defective link in a network by a linear combination of the elements of the base, in the case of polynomial membership. Given a graph G with a link set denoted E, and a vertex set denoted X,

Let a simple graph

$$G = (X, E)$$

be given together with a specified sub-graph G'= (X',E')

and a specified node set X' contained in X –X', we could seek all the maximal cliques of G of G' which are not contained in the set denoted S (X_1) of neighbors on G, [1], where by **clique** we mean a sub-graph of G that is complete, namely every pair of vertices in the clique is joined by a n edge, or link [2, page 177 to 181].

This technique leads to an algorithm with a horrendous complexity, see references [9], [11]. Thus, the method is not very desirable. Fortunately, examination of available research results reveals the existence of two excellent algorithms, [[6] or [7]. A comparison between the latter two is considered next.

EXAMPLE 2. We continue the discussion of example 1,page 82, by giving details on a basis and show how we can carry us to a conclusion that will obviously depend on our ability to determine a base by computer, rather than by inspection as we did in Figure1on page 88.

6. CONCLUSION.

The difference between the two works is that one has a lower complexity than the other does, and in personal preference.

Their use, on networks that are labeled and thus yield a feasible technique replacing a failed link, or node by a linear combination of the elements of the base.

Can we use polynomial expressions as labels? Are they fuzzy numbers [MOF11]? Although published, it is good exercise to repeat the proof here because of its simplicity and because at first, we may doubt it because their sketches tend to oscillate rather wildly. However, **we will show that for each polynomial there exists an interval on which the condition for fuzzy convexity is satisfied. To prove this claim,** we argue that since a polynomial is a linear combination of power functions that are continuous.Therefore, any polynomial P (x) is a continuous function. Furthermore, on any closed interval, it has a maximum, say M, and a minimum, say m. Let M=P (u) and m=P (v). Thus, on the interval [u, v] we have P (s) for all elements s in this interval [u, v] we have that it satisfies the inequality:

$$P (s) \leq \min [P (u),P (v)] \dots\dots\dots\dots\dots (0)$$

Finally, note that each polynomial can be expressed as the vector product of a row vector consisting of all the coefficients of a polynomial with a column vector whose entries are the transpose of the vector below

$$\{1, t\, 2^,\, t\,^3, t\,^4, t\,^5, t\,^6, t\,^n \} \dots (1).$$

Since the polynomials differ at least in their coefficients we can list the entire row vectors in a matrix denoted A. The column vectors as in (1) above also form a matrix we denote as T. Note that its first row consists of entries all equal to 1, the second of al entries equal to t, with third, fourth, fifth, and so on respectively equal to the third, fourth, fifth power of t. Thus, the initial polynomial matrix is expressed as the product of A with T. Its determinant equals the product of their determinants. The value of A can be found with any of the available software. That of T can also be found according to [9, page 123] with a rather low complexity, and easily programmable. It now remains to discuss at least briefly how we compare numerical methods that essentially reach the same goal. And this we discuss in the next section.

7. COMPLEXITY.

How do we decide whenever we find more than one algorithm providing the same goal algorithm, which algorithm is best? We must recall a few basic concepts of the algorithm complexity of a numerical method.

In a nutshell, complexity is a study of the estimated time to execute a given job on a computer. Thus,it normally focuses on the repetition of those operations that are more costly. One such operation is obviously a search. Without giving too many details that are really not pertinent to this research goal. An excellent comprehensive, broad discussion is given by H.Wilf in reference [11]. Suffices here to say that a basic measure is to select a basic operation, then with a probabilistic or a deterministic analysis, we estimate the time that is needed for the completion of the job. Thus, if s denotes the problem size and t is the time needed to run the problem, then either s or t become the independent variables and the complexity is a function of s or of t. As a quick recall, we will distinguish two major types:

(a) The polynomial time estimate, and (b) The exponential time estimate. In the former type, it is required that a function of s be developed where s that stands for the size of the problem, namely a measure of the amount of input.

If the time t is a quadratic function of s, we say that the complexity is of order s^2, and it is denoted by

$O(s^2)$. In the latter type, the function is exponential,in which case it means that it is possible to find four constants denoted c,k,j> 0 and l>1 such that

$$c \bullet k \leq f(s) \leq k \bullet l$$

for all but a finite number of values of s, see reference [11, pages77 and ff.].

Finally, we have the following formal:

DEFINITION 3. a function is said to be of order g (s),

and its notation is:

f (s) is O (g (s)),

If there exists a constant p such that

$$f(s) \leq p \bullet g(s)$$

For all but a finite, and possibly empty, set of nonnegative values of s.

A LAST FEW DETAILS.

How does it all fit together?

In a precious example, we found a basis by inspection, using the definition of independent edges.

Since the fuzzy graph was originally nothing but a labelled graph, and since in linear algebra the first example of linear vector space is that of the set of polynomials of degree N, we will use a basis from the previous example and affix a polynomial label to each edge of the basis. Thus we now proceed to complete the previous example as follows to show how we can determine a linear combination.

EXAMPLE. A set of linearly independent edges we found there were{ [1, 2], [6, 3], [5, 4] }.

Let:

$v_1 = [1, 2], v_2 = [6, 3], v_3 = [5, 4]$

then any linear combination will be using the real numbers r,s,t

$$v = r v_1 + s v_2 + t v_3 \dots\dots\dots\dots \mathbf{(1)}$$

If their respective labels are:

P_1, P_2, P_3

By replacing these labels in the expression (1) above, we have the expression that yields any element v of the linear vector space.

To be more specific, it remains to choose these labels. For instance, we chose the set of polynomials of degree 2 or less such as

$$p_1 = 1, p_2 = 1 + x \text{ and } p_3 = 1 + x + x^2.$$

It remains to show that any quadratic polynomial

$Q(x) = a + bx + cx^2$ is derived uniquely from the three polynomials in the base, namely that

$Q = m p_1 + s p_2 + t p_3$

Where the constants m, t are uniquely found from the given basis, thus, we equate the coefficients of the same powers and this yields:

m+ s +t =a,

s +t = b;

t = c;

The unique solution of the above system is then: t=c,

s= = b – t = b – c, and

m = a – s –t = a – (b – c) - c = a – b +c – c =a – b.

It remains to consider the difficult problem of identifying methods that yields the identification of the maximal spanning set in a network. The selection of such methods depends on the ease with which can be used. Therefore it is necessary to consider the question of complexity of an algorithm.

COMPLEXITY.

How do we decide whenever we find more than one algorithm providing the same goal algorithm, which algorithm is best? We must investigate the algorithm complexity of both numerical methods. In a nutshell, complexity is a study of the estimated time to execute a given job on a computer. Thus,it normally focuses on the repetition of those operations that are more costly. One such operation is obviously a search. Without giving too many details that are really not pertinent to this research goal. An excellent comprehensive discussion is given by H.Wilf in reference [11]. Suffices here to say that a basic measure is to select a basic operation, then with a probabilistic method or a deterministic analysis, we estimate the time that is needed for the completion of the job. Thus, if s denotes the problem size and t is the time needed to run the problem, then either s or t become the independent variables and the complexity is a function of s or of t. As a quick recall, we will distinguish two major types:

(a) The polynomial time estimate, and (b) The exponential time estimate.

In the former type, it is required that a function of s be developed where s that stands for the size of the problem, namely a measure of the amount of input.

If the time t is a quadratic function of s, we say that the complexity is of order s^2, and it is denoted by

$O(s^2)$. In the latter type, the function is exponential, in which case it means that it is possible to find four constants denoted c,k,j> 0 and l>1 such that

$c \bullet k \leq f(s) \leq k \bullet l$

for all but a finite number of values of s [11, pages77and ff.].

Finally, we have the following formal:

Definition. a function is said to be of order g (s),

and its notation is:

$$f(s) \text{ is } O(g(s)),$$

If there exists a constant p such that

$f(s) \leq p \bullet g(s)$

For all but a finite, and possibly empty, set of nonnegative values of s.

8. A last few details. How does it all fit together?

I n a previous example, we found a basis by inspection, using the definition of independent edges.

Since the fuzzy graph was originally nothing but a labelled graph, and since in linear algebra the first example of linear vector space is that of the set of polynomials of degree N, we will use a basis from the previous example and affix a polynomial label to each edge of the basis. Thus we now proceed to complete the previous example as follows to show how we can determine a linear combination.

EXAMPLE. A set of linearly independent edges we had found there were{ [1,2], [6,3], [5,4] }.

Let:

$v_1 = [1,2], v_2 = [6,3], v_3 = [5,4]$

then any linear combination will be using the real numbers r,s,t

$$v = r v_1 + s v_2 + t v_3 \dots\dots\dots\dots (1)$$

If their respective labels are:

$$P_1, P_2, P_3$$

By replacing these labels in the expression (1) above, we have the expression that yields any element v of the linear vector space. To be more specific, it remains to choose these labels. For instance, we chose the set of polynomials of degree 2 or less such as

$$p_1 = 1, p_2 = 1+x \text{ and } p_3 = 1+x+x^2.$$

It remains to show that any quadratic polynomial

$Q(x) = a+bx+cx^2$ is derived uniquely from the three polynomials in the base, namely that

$$Q = m\,p_1 + s\,p_2 + t\,p_3$$

Where the constants m, t are uniquely found from the given basis, thus, we equate the coefficients of the same powers and this yields:

$m + s + t = a,$

$s + t = b;$

$t = c;$

The unique solution of the above system is then: $t=c$,

$s = = b - t = b - c$, and

$m = a - s - t = a - (b - c) - c = a - b + c - c = a - b.$

10. Conclusions. The possibility for a variety of applications is enormous. The last example clearly indicates that once a basis is found then we can uniquely determine by usual linear algebra methods the linear combination of the elements of the base we found thus we can replace the defective edge or vertex with such combination.

Whatever algorithm we choose, is ultimately a matter of personal choice, although few would argue that one with the lowest complexity is the best choice. For other examples of computation of the complexity of a problem we suggest checking the examples in reference [1, pages 75-81].

REFERENCES

for this application of fuzzy graphs to network repairs:

[1] Carre'B. "Graph and Networks, Oxford University Press, London 1979.

[2] — Klir G.&Yuan Bo Fuzzy Sets and fuzzy Logic,Theory and Applications, Prentice Hall,1985.

[3] —McAllister L.N."*A Simulation Technique under Uncertainty*" Proceedings of the 1992 NAFIPS Conference, Puerto Vallarta, Mexico, pages 555-563.

[4] —McAllister L.N., *Fuzzy Intersection Graphs* ",International Journal of Computers and Mathematics with Applications, volume 15, no.10,pages 871-886, 1988.

[5] —McAllister L.N. "*Finding a Base for a Vector Space of Polynomials*", ", 1997 NAFIPS Proceedings,Syracuse, N.Y. U.S.A.., pp283-285. (Isik&Cross, eds.)

[6] —Stefanidis P., Paplinski A.P.,Gibbard M.J. "'*Numerical Operations with Plolynomial Matrices*" Lectures Notes in Control and Information Sciences, (Thomas M.M &Wyner A.Eds.), Springer Verlag, no. 171, 1992.

[7] Tarjan R.E.&Trojanoski A.E"*Finding a Maximum Independent Set*", SIAM J.Comput., vol.6, pages 537-546.1977.

[8] TsukiyamaS.& Ide M.&Shirakawa I."*A New Algorithm for Generating All the Maximal Independent Sets*", SIAM J.Comput., 6, pages 505-517, 1977.

[9] Van Loan C.F. & Golub G.] Matrix Computations, John Hopkins UniversityPress, 1983.

[10] — Rosenfeld A. "*Fuzzy Digital Topology*", Info. and Control, 40, pages 76-87, 1073.

[11] — Wilf H.S. Algorithm and Complexity, Prentice Hall, 1987.

[13] —RomanS. Linear Algebra with Applications, 2nd edition, HBJ, 1988.

Continuing **to provide some answers** to the exercises, we have:

#4. By definition of the recursion we have that the definite integrals over [0,] for $0 \leq x < \infty$,are $f_2(x) = a \int e^{-at} e^{-a(x-t)} dt = a \int e^{-2at+at} =$

$a \int e^{(-at)} dt = a [(-1/a) e^{[-at]} + constant. = -e^{[-at]} + constant.$

To construct an explicit expression: we note that each number denoted K_1 and K_2 have an explicit expression as above what about $K_n = \{x, f_n(x)\}$?

First of all, note that for all $n \geq 2$, we shall find by integration that

$f_n(x) = a^n x^{(n-1)} e^{-ax}] / (n-1)!$

Now, if we assume that the latter holds for $(n-1)$, then we compute $f_n(x)$ according to step3.,then we have precisely the above.

The advantage is that now we can easily find the absolute maximum denoted $M_n = [a (n-1)^{(n-1)}] / [(n-1)! e^{(n-1)}] \ldots\ldots$

And this holds at $x = [n-1]/a$;

using the good old first and second derivative methods.

#6-$|A|= 0.5+0.2+0.4+1= 2.5$;

(ii) $|B| = \sum_{n=0,\ldots,10} [n/(n+1)] = 7.975.$

#7.-. They all are except (iv) if we request that the support be bounded.

#8. (Ii) The domain is R=1-x/100; (ii) The range equals 1-1/y; (iii) R is not one-to-one therefore there is no inverse; it is not symmetric because R (x, y) ≠R (y, x); and not reflexive either because R (x, x) ≠1.

#9. Prove that for each polynomial with real coefficients, there exists ant interval on which the condition of fuzzy convexity holds.

#10. Is it true that if fuzzy convexity holds than convexity in the ordinary sense holds too? Why or why not?

13. INVOLUTIVE means that f (f (x)) =x.

thus we have

g (x) =1-x, therefore g (g (x)) = 1- (1-x) =1- 1+x=cx; q.e.d.

♦ let $Y_N (x) = \{1 - x^N\}^{(1/N)}$

♦ Since

[Y N] N =1- x N

then

$Y_N (Y N) = \{1 - Y N\} (1/N)$

or we simply have

$Y N (Y N) = [1 - (1 - 1 + x N)] (1/N) = x\, x$,

q.e.d;

♦ If $S_\beta (x) = [1 - x] / [1 + \beta x]$

Then $S_\beta (S) = [1 - S] / [1 + \beta S] =$

♦ $[1 + \beta x + \beta - \beta x] / (1 + \beta) =$

♦ $x (x + 1) / (1 + \beta x)\ (1 + \beta x) / (1 + x) =$

♦ after simplification $= x (x + 1) / (1 + x) = x$, q.e.d.;

10. Fuzzy relations. Just as in crisp set theory, we consider, the cross product of two crisp sets to form a set of ordered pairs. Thus, we can then assign a membership valuation to each pair and form a fuzzy relation denoted as follows:

$$R (f) = \{ (a,b), m (a,b) / a\ e\ A, b\ e\ B, m (a,b)\ e\ [0,1] \}.$$

Where the membership function signifies the strength of the association, rather than the probability of that association. This distinction makes sense because of the semantics. For instance, given any couple of our students, we may discuss the probability of their dating, or discuss the strength of the commitment to each other.

10. Fuzzy Relations. Continuing our discussion of these new sets without a sharp boundary, and following the sequence of concepts as we normally develop in set theory, we shall develop the concept of relations between two ordinary crisp sets denoted respectively A and B. Normally, we define the cross product, AxB, as the set of ordered pairs { (a,b), where {a ε A and b ε B}.

How do we fuzzify these? Note that '*fuzzify*" and the like will be expressions that are used to mean that we consider an ordinary crisp case and render it in imprecise fuzzified form. And why should we? This will be made clear as we go along. For example in ordinary set theory, given a collection of terms and a property, we obtain a set that associates to each element either **"1"** to mean that it belongs to the set or **"0"** to mean that it does not belong. Sometimes, we are not able to verify the fact whether or not the property holds; in which case it is not possible to use the characteristic function. Therefore, multivalued logics were invented: In which case, we assume to know that the property holds with a certain evaluation, often a rational number. In probabilistic logic, this valuation is a probability,. In fuzzy logic, we associate to each element a membership degree expressing the degree at which the property is satisfied. The difference between probabilistic and the fuzzy logic case is colossal, all revolving around the meaning and the semantic.

First of all, we could associate to each pair a membership function denoted m (a,b). Again this function need not be a probability because it denotes, not an event, as much as the strength of the association between the two elements. Just as we would when making a distinction between the chance of two individuals dating, and the quality of commitment to one another between these two individuals.

If we wish to give a pictorial representation of a n ordinary set relation, we obtain a graph

$$G = (V,E) \dots\dots\dots\dots\dots\dots\dots\dots\dots\dots\dots \textbf{(5)}$$

where E is contained in the cross product of the set A with the set B where any element of E is called an edge, or link set and V is the set of vertices, or nodes.

Note that in the fuzzy case, we will call the pictorial representation of a fuzzy relation a fuzzy graph..

"Fuzzy Graphs".

Thus, typically a fuzzy graph is a labelled graph. Or it just was considered as depicting a fuzzy relation. Yet, fuzzy graphs have merits and interest in their own right. They it lend themselves to a variety of applications, in which case the membership function may denote the quality of connection between any two points. [AA6] making it possible to devise techniques to compute the fuzziness of information within a network.

Much research has been done on fuzzy graphs beginning from the original work by Rosenfeld,Koczy, Mizumoto, McAllister [AP9,10,11].

For example, we could examine more closely another option.

1. Background.

The original definition of a fuzzy graph was that it simply provided us with a pictorial representation of a fuzzy relation. What is the original definition of a fuzzy graph?

Given two ordinary set A and B, we form an ordinary relation by considering their cross product, [10].

If to each element (a, b) in A x B, we associate a membership function m (a, b), then we form the fuzzy set

$$(\{ (a, b); 0 \le m (a, b) \le 1\} \dots \textbf{(1)}$$

Some, realizing that since the values of m (a, b) are from the unit interval, immediately claimed that the membership was a probability value.

This way, the misunderstanding escalated into a controversy that is promptly resolved after we note that the difference between membership grade and the probability concept is quite great.

Basically the difference is explained by an example emphasizing the distinction in semantic.

The function m (a, b) quantifies the grade of membership of the pair to the fuzzy set. For example suppose we say '"the probability that Matt and Jill are dating is 0.9 ".

When we say "*The commitment of Matt to Jill is 0.9*" the latter numerical value portrays an assessment of the seriousness of the commitment between the two individuals, not the measure of the probability of their dating.

Thus to the pair (Matt, Jill) we associate the membership grade of 0.9

Thus a fuzzy graph depicts the fuzzy set as in the notation (1) above, namely we have

$$G (f) = \{ (a, b); m (a, b) ¤ [0, 1]) \}.... (2)$$

Where (f) reminds us that it is fuzzy.

3. A New Definition of Fuzzy Graph. The new definition states that G (f) as a pair of linear vector spaces. It remains to explain how it is possible to define linear independence for the edge set and for a vertex set of a graph.

Consider a widely accepted definition [1, chapter5]:

Definition 1. We say that a **set of vertices V**, of *nodes*, **is linearly independent** if no two vertices of V are joined by a link, or edge.

Definition2. A **set of edges E,** or of links of a graph is said to be **linearly independent** if no two links in the set have a vertex in common.

Example [FG-Figure]. Let the set links E= { [1, 2], [1, 3], [2, 4], [1, 6], [6, 4], [5, 4], [3, 4], [6, 3] } where the vertices are shown in the figure 1, see also [2, p.175].

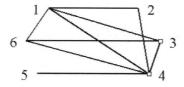

FG-Figure 1. The above figure shows a graph whose vertex set V={1, 2, 3, 4, 5, 6} has maximal independent sets, or a base, that are {1, 5}, {2, 3, 5}, and {2, 5, 6}, {4}, see [1, page 175]. Note that the links [1,2], [6,3], and [5,4] have no vertex in common, thus they are linearly independent.

However, a search based on directly on the definition would be very time-consuming, therefore not feasible for any graph of a considerable size.

4. The Advantage of the New Definition of fuzzy graph is that if, for each vector space, we can find a base, then any link, or vertex that fails, can be expressed as a linear combination of the elements of its base, or of its maximal spanning tree.

Is there a contradiction between the former definition of a fuzzy graph as a pair of labeled graphs and the current as a pair of vector spaces? The question has been answered with great care in a negative way in [3].

To be more specific, let

$$\{E_1, E_2, E_3, E_4, e_5, E_6, E_n\} .. (3)$$

be a base For the crisp link set E we and by any method as in [5,6,7], then our preceding discussion yields the following

Corollary. The set of links forming a base for the crisp case, provides also a base for the fuzzy link set.

Proof. In other words, we must be able to argue that if (3) forms a base of the support crisp link set E of the fuzzy set E (f) where again (f) denotes the fuzzy case then the choices:

$$p_1 = 1, p_2 = 1+x \text{ and } p_3 = 1+x+x^2.$$

It remains to show that any quadratic polynomial

$Q(x) = c+bx+ax^2$ is derived uniquely from the three polynomials in the base, namely that

$Q = m p_1 + s p_2 + t p_3$

Where the constants m, t are uniquely found from the given basis, thus, we equate the coefficients of the same powers and this yields:

m+ s +t =a,

s +t = b;

t = c;

The unique solution of the above system is then: t=c,

s= = b – t = b – c, and

m = a – s –t = a – (b – c) - c = a – b +c – c =a – b.

Conclusions. The potential for meaningful applications is great as the last example clearly indicates that once a basis is found then we can uniquely determine by usual linear algebra methods the linear combination of the elements of the base we found thus we can replace the defective edge or vertex with such combination, [15].

Whatever algorithm we choose, is ultimately a matter of personal choice, although few would argue that one with the lowest complexity is the best choice. For other examples of computation of the complexity of a problem we suggest checking the examples in [1, pages 75-81].

REFERENCES

[1] - Carre's. "Graph and Networks, Oxford University Press, London 1979;

[2] — Klir G.&Yuan Bo Fuzzy Sets and fuzzy Logic,Theory and Applications, Prentice Hall,1985;

[3] —McAllister L.N."*A Simulation Technique under Uncertainty*" Proceedings of the 1992 NAFIPS Conference, Puerto Vallarta, Mexico, pages 555-563.;

[4] —McAllister L.N., *Fuzzy Intersection Graphs* ", International Journal of Computers and Mathematics with Applications, volume 15, no.10,pages 871-886, 1988.;

[5] —McAllister L.N. "*Finding a Base for a Vector Space of Polynomials*", ", 1997 NAFIPS Proceedings,Syracuse, N.Y. U.S.A.., pp283-285. (Isik&Cross, eds.)

[6] Stefanidis P., Paplinski A.P.,Gibbard M.J. '"*Numerical Operations with Plolynomial Matrices*" Lectures Notes in Control and Information Sciences, (Thomas M.M &Wyner A.Eds.), Springer Verlag, no. 171, 1992;

[7] Tarjan R.E.&Trojanoski A.E"*Finding a Maximum Independent Set*", SIAM J.Comput., vol.6, pages 537-546.1977;

[8] TsukiyamaS.& Ide M.&Shirakawa I."*A New Algorithm for Generating All the Maximal Independent Sets*"., SIAM J.Comput., 6, pages 505-517, 1977.

[9] Van Loan C.F. & Golub G.] Matrix Computations, John Hopkins University Press, 1983.

[10] —Rosenfeld A. "*Fuzzy Digital Topology*", Info. and Control, 40, pages 76-87, 1073.

[11] — Wilf H.S. Algorithm and Complexity, Prentice Hall, 1987;

[12].Mc Allister L.N. "*Inexact Graphs: Their Role in Some Expert Systems* ", PROC. Of the 1-st CONF.on applications of Artificial Intelligence, Denver,CO,pp.505-509,December 1983;

[13].McAllisterL.N. "*A Measure of the Quality of Connection in a Directed Graph*", The Analysis of Fuzzy Information, by J.C.Bezdek, CRC Press,chapter9, 1986;

[14].McAllister L.N."*Simulation of Control for Automated Manufacturing*" PROCEED. of the CONF. On Artificial Intelligence and Advanced Computer Techniques, Long Beach,CA,pp.83-88,1985.

[15] McAllister L.N. "*Fuzzy Graphs and Network Repairs*"', Intern.Journal of Computers in Mathematics, vol.80, no.9, pp.1-6, Sep.2003.

CHAPTER 12: SOFT COMPUTING

Benjoe A.Juliano [AP25, Chapter 7]

The conversion of perceptions into measurement is unfeasible, unrealistic or counter productive"

Soft Computing is the name of a new field that is also referred to *as "computing with words rather than numbers.*

We say that it is the way of the future. Why? How many times we are given a questionnaire where the response is requested in the form of a number between 0 and 10 with 0 denoting disapproval, and 10 meaning complete approval.? As we have previously remarked, we believe that the response, in the form of a choice of a number, is most often a random selection, and seldom consistent. If we were asked to choose an adjective to describe our response, chances are that it would take a longer time. However, ultimately, the response is most likely to be more valid and consistent because it is rationalized. Processing such responses in questionnaires might still involve some quantification, as it is done often. However, the best procedure is to device more realistic procedures that approximate, but do not replace the response.

In this fashion, fuzzy logic is undoubtedly the most suitable tool. As it was stated on page 13, fuzzy logic provides an approximation to classical logic just as numerical analysis provides approximate solutions to problems for which we cannot find an exact solution easily. Thus, we now have a tool that we can rely on to manipulate uncertainty that stems from approximations we can suitably control. Modeling techniques are devised accordingly, see [U10, **U15**].

Soft Computing, or computing with words, is a fascinating new way of investigating problems that are related to know ledge representation, symbolic reasoning, natural language, and intelligent control using the concepts of fuzzy sets and fuzzy logic, see for example [AP18]. Research in fuzzy logic continues to grow in its technical and theoretical areas. Much work is currently done so much so that researchers have dubbed, according to my friend Paul Wang, *"The twenty-first century as the Century of Semantics", see* Reference [AP22]. His book is timely. It responds to a need and demand for a book on this evolving field of computing with words. His book consists of a collection of chapters, each written by a leading researcher within the fuzzy community. It provides a complete overview on its theory and potential applications. A view of computing with words with semiotics and intelligent systems is also presented. Thus, to conclude, it is an essential source of information. A journal is published in Spain at the University of Barcelona. A center called BISC [short for Berkeley Initiative on Soft Computing],exists that promotes research in the field under the direction of L. A.Zadeh, who first proposed and led the world research in the theory of fuzzy sets, then in fuzzy logic, and now in soft computing. Why is it that computing with words rather than numbers is important? Imagine for a moment of having to make some sense out of sketchy descriptions of a crime perpetrator, something like:

♦   Male of medium height and build;

♦   Swarthy complexion;

♦   Fairly tall;

♦   A grayish-blond beard;

♦   Not black, not white complexion;

As outlandish as this case might appear, in reality, police is faced with these situations any time a crime occurs and they try to interview witnesses and must make sense of whatever information has been collected. **Soft Computing** can help draw some sense out of the above descriptions, much better that any numerical assessment. This is a quite compelling reason why we must consider this field as the way to the future to exploit its enormous capabilities. Why do I refer to it as the way of the future? There is a simple compelling reason. Consider the vagueness of the assessments that were listed above as examples. Why pinpoint a selection because of some probability value? When this value is often hypothetical and uncertain? To each assessment there corresponds an infinity of possibilities.Why restrict ourselves? If to each assessment above we associate a fuzzy number then the infinity of possibilities still holds unrestricted, and it keeps its characteristic property. Therefore the interpretation of each assessment holds intact. If such application, at this time. Seems farfetched, it is not however. Etienne Kerre of the Applied Mathematics Seminar at the University of Gent, Belgium has been already involved with applications rather close to this kind in police investigations. Another important application occurs when we wish to analyze approximate reasoning that allows a human to make sense of garbled speech or to fill in imprecise, incomplete information.

This methodology relies heavily on the ability of representing sentences by fuzzy numbers, as we point out in the following two examples. Where, in the first case, we derive the overall assessment of the performance of a skater during a performance of a program lasting a total of 100 minutes, and the second more easily lends to the representation of vague statements on a person's age.

EXAMPLE 1. The representation of the performance of a skater

♦   **A trapezoidal fuzzy number**; T [0, 20, 40, 100]

♦   **At times t=0, 20, 40,100, the valuations are crisp, either 0 (poor) or 1 (excellent).** Note that the notation that was agreed upon by the fuzzy community requires that the middle two numbers are associated to a valuation equal to 1, but the two outer numbers have a valuation equal to zero, providing a total of the four vertices of the trapezoid!
Its vertices are:
• A (0,0), B (20, 1), C (40, 1), and D (100, 0).
Note the **unexpected BONUS**:

Since, the line BC is on y=1, and the line AD on y=0. Note that we can also find the equation of the two sloping lines, i.e. AB, and CD. Thus, it is possible to compute the performance assessment at any time between the selected intervals. How? We simply find the equation of the line for AB, or for CD, and then evaluate the y-coordinate for the chosen value of t. And also, if necessary, between t=40, and t=100. Once again we have established a correspondence between a geometrical shape and a sentence The meaning of the trapezoid is the representation of the overall performance of a skater between t=0, and t=100. Once we have the ability to rank fuzzy numbers [and we recall that an easy algorithm was written and published by Buckley in IJFSS, see reference [24], where he has described an algorithm that has as output the ranking the trapezoidal numbers.Note the im -portance of this approach that does not require guessing overall estimates as values that do not record consistently the performance of any athlete, thus the figure below corresponds to the evaluation of the skater's performance.

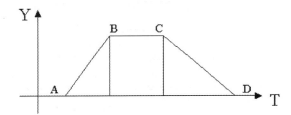

EXAMPLE 2. The representation of **"mid-sixty"**

In this case, it is easier to use a different shape, namely we have:

2-a Triangular Fuzzy Number T [60, 65, 65, 70].

Note that the two middle numbers are the same, and both have the same valuation [i.e. one], since it is a vertex of the triangle; likewise for the outer values [i.e. zero]; Recall that there are only three significant vertices!

Its vertices are:

A (60,0); B (65,1); and C (70,0),

where the valuations correspond reasonably to the fact 65 is certainly **midsixty**, (thus, its valuation ought to equal 1),

while neither 60 or 70 do (thus, its valuation=0).he figure below thus corresponds to the representation of a vague sentence such as

"Mrs. Dara is midsixty"

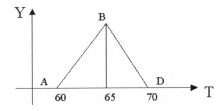

CONCLUSION

The goal of this concise introduction to fuzzy logic was to point out just its versatility and simplicity, so that it ought not to be surprising that during the last few decades, it has achieved great popularity in its use in a variety of applications. It resembles humans' approximate reasoning, or the humans' ability of making sense out of sloppy speech or of filling gaps when confronted with incomplete information.

To fill in the incomplete information of this introduction, we urge the interested reader to scan the bibliography that includes a special set of references that are dedicated to a variety of topics. Soft Computing is already investigated extensively by fuzzy researchers with excellent results that apply to machine intelligence, see references [25, 26].

Of course other attempts have been made, For example, see [MOF17], where the approach to reasoning is based on interval-valued fuzzy sets, or as in Yao's study, a comparison is made with probabilistic case [MOF18], rather than the fuzzy case.case

ABOUT THE AUTHOR: BRIEF VITA HIGHLIGHTS

Luisa Nicosia McAllister received a Doctor in Mathematical Sciences from the University of Rome, Italy in 1957. She then went to the University California at Berkeley where she specialized by studying Numerical Analysis and Computer Science under the direction of professor R.De Vogelaere, and was hired by him as his Research Assistant After moving to the east coast, she taught applied Numerical Analysis and Real Analysis for the University of Delaware extension at the Aberdeen Ballistic Laboratory; and Linear Algebra, and Advanced Calculus for the University of Maryland exten sion at the Aberdeen Ballistic Laboratory. She also taught, 1964-65, at Towson State University. Finally, moving to Bethlehem, in 1965, she joined the Mathematics Department of Moravian College. She was promoted to professor in 1981, and retired for personal reasons in March 2001. She is an American citizen and is married with three adult children. She moved with her husband Gregory to a retirement community in Nazateth, PA.

She is the author of over 30 research papers in fuzzy graphs and their applications; She served: **1**—on the executive committee of the EPADEL, the local chapter of the MAA; **2**—on the board of directors of NAFIPS (short for North-American Fuzzy Information Processing Society) serving as its bulletin editor from1979 to 1993; **3**—on the executive committee of EPADEL section of the MAA; **3**—on the Lehigh Valley section of IEEE.; **4**—**Currently** she serves as the chair of the information committee at BISC, Comp.Sci. Dept, University of CA, Berkeley.; and on the advisory board of AIM.

Among her many awards, she includes two Fulbright fellowships, one personal NSF grant that allowed her to study Graph Theory at the University of Michigan, Ann Arbor under the direction of Professor F. Harary, and Fuzzy Logic and Expert Systems at the University of California, Berkeley under the direction of Professor L.A. Zadeh, and to present a research paper at the Polish Academy of Sciences in Warsaw, Poland in november 1986;

She was included in two collective NSF grant that allowed her to present her research results in 1984 at the Bejin Electrical Power Institute and at the first IFSA meeting in Hawai, and in 1987 at the university of canton, China; at the second IFSA meeting in Tokyo, Japan;

BOOK MERITORY HIGHLIGHTS

It provides a brief introduction to a few basic concepts.

Its value consists in addressing the interest of a reader who has special interests by providing a selection of authoritative research books and papers on different fields, such as fuzzy mathematics, a wide range of applications, new theories in modeling uncertainty.